Novel Nanocomposites

Novel Nanocomposites

Optical, Electrical, Mechanical and Surface Related Properties

Editors

Mirela Suchea
Emmanouel Koudoumas
Petronela Pascariu

MDPI • Basel • Beijing • Wuhan • Barcelona • Belgrade • Manchester • Tokyo • Cluj • Tianjin

Editors

Mirela Suchea
National Institute for Research
and Development in
Microtechnologies-IMT
Bucharest
Romania
Center of Materials Technology
and Photonics, Hellenic
Mediterranean University
Greece

Emmanouel Koudoumas
Center of Materials Technology
and Photonics, Hellenic
Mediterranean University
(formerly Technological
Educational Institute of Crete)
Greece

Petronela Pascariu
"Petru Poni" Institute of
Macromolecular Chemistry
Romania

Editorial Office
MDPI
St. Alban-Anlage 66
4052 Basel, Switzerland

This is a reprint of articles from the Special Issue published online in the open access journal *Nanomaterials* (ISSN 2079-4991) (available at: https://www.mdpi.com/journal/nanomaterials/ special_issues/novel_nanocomposites_properties).

For citation purposes, cite each article independently as indicated on the article page online and as indicated below:

LastName, A.A.; LastName, B.B.; LastName, C.C. Article Title. *Journal Name* **Year**, *Volume Number*, Page Range.

ISBN 978-3-0365-2247-0 (Hbk)
ISBN 978-3-0365-2248-7 (PDF)

Cover image courtesy of Mirela Petruta Suchea and Ioan Valentin Tudose (NanoCandies-ZnO).

Contents

About the Editors

Mirela Suchea received her License in Physics–Biophysics in 1999 from the University of Bucharest, Romania, and her PhD in Physical Chemistry from the Chemistry Department of the University of Crete in 2009. She started her research career in 2002, and currently works as senior Scientific Researcher I at IMT Bucharest in Romania, and in Greece as an associated researcher at the Smart Functional Materials Group at CEMATEP, Hellenic Mediterranean University (HMU). Dr. Mirela Suchea has dedicated her energies to research in the field of nanostructured materials synthesis, growth and applications. She works in the materials science field and nanotechnology, and is the author or coauthor of more than 70 journal articles (over 1250 citations) and book chapters. Her work is broadly interdisciplinary between materials science, physics, chemistry, electrochemistry, instruments and instrumentations, and technology having a H-index 17 according to web of science. (for online version please link to https://publons.com/researcher/1266831/mirela-suchea). Beside her scientific activity, she is a NanoArt passionate artist. She is *The Academy of NanoArt* director for Europe region.

Emmanouel Koudoumas is a Professor in the Department of Electrical and Computer Engineering at the Hellenic Mediterranean University, Greece. His research efforts concern the use of simple, low cost and environmental friendly techniques for the growth of thin films, nanostructure layers, nanomaterials, nanocomposites and suspensions, based on pure or doped metal oxides, metals or carbon allotropes, exhibiting properties suitable for control light transmission, photo-catalytic/antifouling/antibacterial action, self-cleaning windows, Li ion batteries, electromagnetic shielding, nanodielectrics, food packaging, etc. He has 140 publications in peer-review journals, with 3200 citations and an h-index of 33, while he has been involved in a large number of funded projects as coordinator or member of the research team.

Petronela Pascariu received her PhD in Physics at Al. I. Cuza University Iasi Romania in the area of functional multilayered nanostructures for spintronic applications (September 2011). Currently, she is senior Researcher (CSIII) at "Petru Poni" Institute of Macromolecular Chemistry. She has dedicated her energies to the design, synthesis and investigation of nanoparticles, thin films and multilayered nanowires and metal oxide ceramic nanofibers based on ZnO and TiO_2 doped with Ni, Co, Ag, La, Er, Sm, Mo, etc., composites, polymer/inorganic nanoparticles nanostructures, with specific optical, electrical, photocatalytic and magnetic properties for use in various modern applications (photocatalysis, sensors, electronics and optoelectronics, electrochemical supercapacitors, etc.), which are obtained by electrochemical and electrospinning methods, respectively.

Preface to "Novel Nanocomposites"

The development of novel nanocomposite materials with enhanced physical and chemical properties is one of the most important emergent topics in recent years. Since the development of affordable and commercially available nanomaterials, nanocomposites have become a highly desirable product. The development of novel nanocomposites with enhanced optical properties became of real interest for domains such as data transmission, sensors, nonlinear optics devices, etc. Electric/dielectric nanocomposites are primary areas of focus within the research for novel supercapacitors or other renewable energy applications; nanocomposites with specific surface properties have become very attractive for antistatic, antibacterial, photocatalytic, etc., functional surface applications while enhanced mechanical properties are desirable for a plethora of needs in transportation, home appliances, architectural applications, etc. Having multiple levels of functionality simultaneously is possible using nanocomposite materials, by achieving a synergistic effect of nano-components/matrix properties. The present Special Issue, "Novel Nanocomposites: Optical, Electrical, Mechanical and Surface Related Properties", presents a collection of eight original, high-quality research papers covering a broad range of subjects, from nanocomposite synthesis/fabrication to the design and characterization of various nanocomposites materials with enhanced optical, electrical, mechanical, and surface-related properties as well as practical applications.

Mirela Suchea, Emmanouel Koudoumas, Petronela Pascariu
Editors

nanomaterials

Article

Obtaining Nanostructured ZnO onto Si Coatings for Optoelectronic Applications via Eco-Friendly Chemical Preparation Routes

Mirela Petruta Suchea [1,2,*], Evangelia Petromichelaki [3,4], Cosmin Romanitan [2], Maria Androulidaki [3], Alexandra Manousaki [3], Zacharias Viskadourakis [3], Rabia Ikram [4,5], Petronela Pascariu [6,*] and George Kenanakis [3,*]

[1] Center of Materials Technology and Photonics, Hellenic Mediterranean University, 71410 Heraklion, Greece
[2] National Institute for Research and Development in Microtechnologies (IMT-Bucharest), 023573 Bucharest, Romania; cosmin.romanitan@imt.ro
[3] Institute of Electronic Structure and Laser, Foundation for Research & Technology-Hellas, N. Plastira 100, 70013 Heraklion, Greece; epetromichelaki@physics.uoc.gr (E.P.); pyrhnas@physics.uoc.gr (M.A.); manousa@iesl.forth.gr (A.M.); zach@iesl.forth.gr (Z.V.)
[4] Physics Department, University of Crete, 71003 Heraklion, Greece; raab@um.edu.my
[5] Department of Chemical Engineering, University of Malaya, Kuala Lumpur 50603, Malaysia
[6] "Petru Poni" Institute of Macromolecular Chemistry, 700487 Iasi, Romania
* Correspondence: mirasuchea@hmu.gr or mira.suchea@imt.ro (M.P.S.); dorneanu.petronela@icmpp.ro (P.P.); gkenanak@iesl.forth.gr (G.K.)

Citation: Suchea, M.P.; Petromichelaki, E.; Romanitan, C.; Androulidaki, M.; Manousaki, A.; Viskadourakis, Z.; Ikram, R.; Pascariu, P.; Kenanakis, G. Obtaining Nanostructured ZnO onto Si Coatings for Optoelectronic Applications via Eco-Friendly Chemical Preparation Routes. *Nanomaterials* **2021**, *11*, 2490. https://doi.org/10.3390/nano11102490

Academic Editors: Fabien Grasset and Andres Castellanos-Gomez

Received: 18 August 2021
Accepted: 13 September 2021
Published: 24 September 2021

Publisher's Note: MDPI stays neutral with regard to jurisdictional claims in published maps and institutional affiliations.

Abstract: Although the research on zinc oxide (ZnO) has a very long history and its applications are almost countless as the publications on this subject are extensive, this semiconductor is still full of resources and continues to offer very interesting results worth publishing or warrants further investigation. The recent years are marked by the development of novel green chemical synthesis routes for semiconductor fabrication in order to reduce the environmental impacts associated with synthesis on one hand and to inhibit/suppress the toxicity and hazards at the end of their lifecycle on the other hand. In this context, this study focused on the development of various kinds of nanostructured ZnO onto Si substrates via chemical route synthesis using both classic solvents and some usual non-toxic beverages to substitute the expensive high purity reagents acquired from specialized providers. To our knowledge, this represents the first systematic study involving common beverages as reagents in order to obtain ZnO coatings onto Si for optoelectronic applications by the Aqueous Chemical Growth (ACG) technique. Moreover, the present study offers comparative information on obtaining nanostructured ZnO coatings with a large variety of bulk and surface morphologies consisting of crystalline nanostructures. It was revealed from X-ray diffraction analysis via Williamson–Hall plots that the resulting wurtzite ZnO has a large crystallite size and small lattice strain. These morphological features resulted in good optical properties, as proved by photoluminescence (PL) measurements even at room temperature (295 K). Good optical properties could be ascribed to complex surface structuring and large surface-to-volume ratios.

Keywords: zinc oxide; green synthesis; nanostructured; optical properties; optoelectronics

1. Introduction

In optoelectronics, ZnO is commonly used as a polycrystalline nanostructured coating, thin film, or single crystals. As the optical character of the material depends on its structure, size, and shape of building blocks and the resulting surface structure that is a direct consequence of the preparation process, the main goal of this work is to compare the optical properties of various kinds of nanostructured ZnO coatings resulting from different kinds of syntheses, including also a few green routes. The comparison was done by evaluating PL spectral properties of ZnO layers grown onto Si substrates and relating them to the specific bulk and surface structure via respective synthesis routes.

The wurtzite-type structure of ZnO is the most thermodynamically stable form of this semiconductor and exhibits a wide bandgap at around 3.37 eV [1,2]. Compared to gallium nitride, which is a semiconductor with a similar energy gap, ZnO possesses strong polar binding, large exciton binding energy (60 meV), and relatively high longitudinal optical (LO) phonon energy of 72 meV [2]. Due to the presence of oxygen and zinc atoms interlaced along the c-axis, wurtzite-type ZnO is polar in this direction, thus gaining piezoelectric properties. Although the as grown ZnO exhibits n-type conductivity [1,2], the development of methods of n-doped and p-doped ZnO thin film deposition is still of vivid interest. Many deposition techniques were used during ZnO thin films history: physical growth methods, chemical methods, and combinations of them. Among chemical techniques, aqueous chemical growth, sol-gel, dip coating, and biosynthesis are the most common ones [3–14]. Due to the aforementioned properties, the ZnO and its composite materials are considered to be great candidates for light-emitting and laser devices, which cover the spectral range from green over the blue to the near-UV. In addition, sensors, optical filters, and nonlinear optics devices could be obtained by virtue of the remarkable ZnO properties [2]. Undoped ZnO layers exhibit high electron concentration and are quite insensitive to visible light; thus, they are also suitable as a semiconductor in thin-film transistors [2] or as transparent conductive electrodes for various applications [2]. Compared to ITO films, ZnO layers are chemically stable, can be safely fabricated at low-cost, and are harmless for living creatures [2]. Piezoelectric and ferroelectric properties of ZnO films allow them be used also in acoustoelectronic, acousto-optic, and memory devices [1,2]. Although the research on ZnO has a very long history and its applications are almost countless as the publications on this subject are extensive, this semiconductor is still full of resources and continues to offer very interesting results worth publishing or warrants further investigation [15,16]. The recent years are marked by the development of novel green chemical synthesis routes for semiconductors fabrication in order to reduce the environmental impact of fabrication on one hand and reduce the toxicity and hazards at the end of their lifecycle on the other. Green chemical syntheses are targeting the design of chemical products and processes that reduce or eliminate the use or generation of hazardous substances. Green chemistry applies across the life cycle of a chemical product, including its design, manufacture, use, and ultimate disposal. Reducing or eliminating one or more high purity solvents or reagents from a process result in a whole chain of substantial energy consumption, reduced use of hazardous and toxic materials consumption used in the synthesis and purification of the respective solvent or reagent, and generates long term positive effects on reducing the environmental impact of chemical processing. As an example, ethanol is one of the most used and cheap solvents, and it is synthesized by the hydration of ethylene and by fermentation. In semiconductor fabrication, the highest purity of materials is required; thus, ethanol synthesized by the ethylene hydration process is mostly used. Ethylene is an essential chemical in the petrochemical industry. Ethylene is traditionally produced through the steam cracking of hydrocarbons, and this method remains the predominant method in the industry. In this process, ethanol is produced by a reversible exothermic reaction between ethylene and water vapor. The process consists of three different steps including reaction, recovery, and purification. Ethylene conversion is about 4–25%, and it is recycled. Ethanol selectivity is 98.5 mol%. Phosphoric (V) acid coated onto a solid silicon dioxide has been used mainly as the catalyst. Phosphoric acid is used as a catalyst, and conversion is 4–25%. Acetaldehyde is produced as a byproduct, which can either be sold or further hydrogenated in order to produce ethanol. The unreacted reactants are separated from the vapor mixture of the reactor in a high-pressure separator and then scrubbed with water to dissolve the ethanol. Ethanol–water mixture forms an azeotrope mixture that needs special distillation techniques, which eventually increase the costs. This process is not economically feasible because ethanol is a cheaper chemical than ethylene. Moreover, it is non-renewable if ethylene is produced by hydrocracking petroleum products. Therefore, validating new technologies that may help to reduce the need for chemicals produced via such kinds of chemical process is necessary.

In this context, this study focused on the development of various kinds of nanostructured ZnO onto Si substrates via chemical route synthesis using both classic solvents and some usual non-toxic beverages to substitute the expensive high purity reagents acquired from specialized providers. To our knowledge, this is the first systematic study involving common beverages as reagents in order to fabricate ZnO coatings onto Si for electronic applications by ACG technique. Moreover, the present study offers comparative information on obtaining nanostructured ZnO coatings with a large variety of bulk and surface morphologies consisting of crystalline nanostructures with high crystallinity with wurtzite type structure suitable for applications that require complex surface structuring with a high number of grain boundaries and large surface-to-volume ratios and individual highly crystalline ZnO nanostructures such as nonlinear optics and some optoelectronic applications. All fabricated materials were thoroughly characterized by SEM, EDX, and XRD, and their PL spectroscopic properties were analyzed.

2. Materials and Methods

The general idea of this work was to synthesize ZnO by using chemical reactants and then try to substitute each one of them with nontoxic and mild reagents, using the Aqueous Chemical Growth (ACG) method.

All experiments were performed at relatively low temperatures. During the synthesis of nanostructured ZnO coatings onto Si substrates, the reaction mechanism is the following.

$$Zn^{+2} + 2OH^- \leftrightarrow Zn(OH)_2 \leftrightarrow ZnO + H_2O \tag{R1}$$

In reaction (R1), the source of zinc cations is typically zinc salts, such as zinc nitrate ($Zn(NO_3)_2$), zinc acetate ($Zn(CH_3COO)_2$), or even metallic Zn. The hydroxyl anions can result from the presence of water, alcohol, and other solvents containing hydroxyl radicals. Moreover, OH^- can be formed by the reaction of an amine, such as hexamethylenetetramine (HMTA, $(CH_2)_6N_4$) or even Ammonium carbonate (($(NH_4)_2CO_3$), with water.

The pH value variation during the reaction could not be monitored due to the fact that the high-pressure conditions in the autoclaves do not allow this. The effect of pH on this kind of reaction was studied during an older work [17]. It was observed that as the pH values increase, the precipitation of ZnO nanostructures starts earlier compared to lower pH values, but the crystal quality becomes rather poor. The poor crystal quality for large pH values can be attributed to a higher reaction rate, which can be confirmed by the increase in precipitation rate of the material while increasing pH value. From the SEM characterization, it is obvious that the pH values indeed affect the morphology of the as-grown ZnO nanostructures according to each solvent that was used.

First, pure chemical reactants, $Zn(NO_3)_2$ or $Zn(CH_3COO)_2$ and HMTA, were used in different solvents (water, ethanol, etc.). The samples resulting from these experiments are used as reference samples and for direct comparison with others from the existing literature.

Eco-friendly synthesis was achieved by substituting the chemical reactant of HMTA with ammonium carbonate using as a source of baking ammonia sold in the local food market. Baking ammonia has been listed as "Generally Recognized as Safe" (GRAS) by the US Food and Drug Administration (21CFR184.1137) [18]. Baking ammonia is a mixture of ammonium bicarbonate (NH_4HCO_3) and ammonium carbamate (NH_2COONH_4). It has been used as a primary leavening agent by bakers, before the advent of baking soda and baking powder, because under heating conditions it breaks into carbon dioxide (leavening), ammonia (needs dissipation), and water [19]. The experiments were performed using the same solvents and under the same temperature conditions by using both types of amines.

To our best knowledge, it is the first time that baking ammonia is used for the ACG synthesis of nanostructured ZnO coatings onto Si substrates. Si substrates were chosen because Si technology is still the most important and used in our days for optoelectronic devices. Si was choose to ensure that the proposed method would be compatible with the actual optoelectronic and microelectronic devices technologies and will allow the obtained ZnO materials to be integrated easily with other components in real life applications. Some preliminary studies on the use of baking ammonia to grow photocatalytic ZnO onto PLA 3D printed scaffolds were reported recently by the same research team [20]. After a systematic study of using baking ammonia as the OH^- source, the final step for green synthesized ZnO was to substitute the zinc source. According to the literature, ZnO can be obtained by oxidation of metallic Zn powder [21–25]. Consequently, instead of using $Zn(NO_3)_2$ or $Zn(CH_3COO)_2$, Zn metal powder was used. Various experiments were performed by using HMTA and baking ammonia dissolved in various solvents. Several syntheses were also performed without the use of any type of amine.

2.1. Materials

ZnO nanostructured coatings were grown by the ACG technique onto well-cleaned silicon (100) wafers obtained from SILCHEM Handelsgesellschaft mbH (Freiberg, Germany) using zinc nitrate hexahydrate $Zn(NO_3)_2 \cdot 6H_2O$, $\geq$99.99%; zinc acetate dihydrate $Zn(CH_3COO)_2 \cdot 2H_2O$, $\geq$99.99%; hexamethylenetetramine $(CH_2)_6N_4$, >99.00%; and ethylic alcohol, all purchased from, Sigma-Aldrich/Merck KGaA, Darmstadt, Germany and Zn powder from Fluka. Seelze, Switzerland. All of these were used without any further purification. Si wafers were as follows: single side polished; orientation: (100) $\pm$ 1°; type: p/boron; resistivity: 15.0 Ohmcm; thickness: 400 $\pm$ 25 μm. For the eco-friendly syntheses, nontoxic and daily use reactants were used. Specifically, in addition to water, beverages such as soda water, leavening agents used in traditional puff pastry and cookie recipes such as baking ammonia, mild antiseptics such as hydrogen peroxide (2.8% w/w), and ethyl alcohol based alcoholic beverages such as Raki and Ouzo (Greek products) were used for the synthesis of the nanostructured ZnO.

2.2. Syntheses

1. The amount of 50 mL of equimolar (0.01 M) aqueous solution of $Zn(NO_3)_2 \cdot 6H_2O$ and HMTA was placed in a common laboratory oven preheated at a specific temperature (95 °C or 195 °C) for 2 h, followed by washing with deionized water to remove residual contaminants.
2. The amount of 50 mL of equimolar (0.01 M) aqueous solution of $Zn(CH_3COO)_2 \cdot 2H_2O$ and HMTA was placed in a common laboratory oven preheated at a specific temperature (95 °C or 195 °C) for 2 h, followed by washing with deionized water to remove residual contaminants.
3. Replacing the water with ethanol, raki, and Ouzo for 1 and 2 synthesis conditions.
4. Replacing HMTA with the nontoxic baking ammonia for 1, 2, and 3 synthesis conditions.
5. Replacing the Zn source with Zn metallic. The amount of 1.8 g metallic Zn powder was dissolved in 35 mL of distilled water and 0.05 g HMTA under continuous stirring, followed by thermal treatment in a laboratory oven at 195 °C for 24 h. Washing occurred after.
6. Replacing water in synthesis 5 conditions with ethanol, soda water, lemon soft drink, and hydrogen peroxide (2.8% w/w).
7. Changing the HMTA amount. Experiments that use 0.15 g instead of 0.05 g HMTA were also performed under the exact same procedure by using the different solvents (water, soda water, lemon soft drink, and hydrogen peroxide).
8. Replacing the HMTA with 0.05 g and 0.15 g nontoxic baking ammonia in synthesis 5 and 6 conditions.

9. Elimination of the amine: 1.8 g metallic Zn powder dissolved in 35 mL of the desired solvent (distilled water, ethanol, soda water, and hydrogen peroxide), followed by thermal treatment in a laboratory oven at 195 °C for 24 h. Washing occurred after.

2.3. Characterization

All the prepared samples were characterized by using Scanning Electron Microscopy (SEM) JSM-6390LV SEM from JEOL equipped with Inka-act Energy Dispersive X-ray Analysis (EDX) system from Oxford and X-ray Diffraction (XRD) using using a Rigaku RINT-2000 X-ray diffractometer. XRD patterns were recorded from 30° to 70° 2θ angles using CuKα1 radiation with a monochromatic wavelength of 1.5405 Å operated at 40 kV and 82 mA. Photoluminescence (PL) spectroscopy using a He-Cd, cw laser at 325 nm with full power 35 mW was also performed. The samples were placed in a high vacuum cryostat, which was cooled down to change the temperature from 300 K to 13 K. The emission spectrum was measured using a very sensitive, LN_2 cooled CCD camera.

3. Results and Discussion

3.1. Growth and Structuring

Scanning electron microscopy SEM micrographs were analyzed in order to examine the morphology of the prepared samples, while energy dispersive X-ray EDX analysis provided their elemental analysis. Their structural properties were assessed by using XRD measurements via a size-strain Williamson–Hall plot. In addition, the optical properties were studied through PL measurements.

Morphology and Structuring onto Si Substrates

SEM images of the samples synthesized via synthesis routes 1, 2, and 3 at 95 °C are presented in Figure 1 for a synthesis temperature of 95 °C.

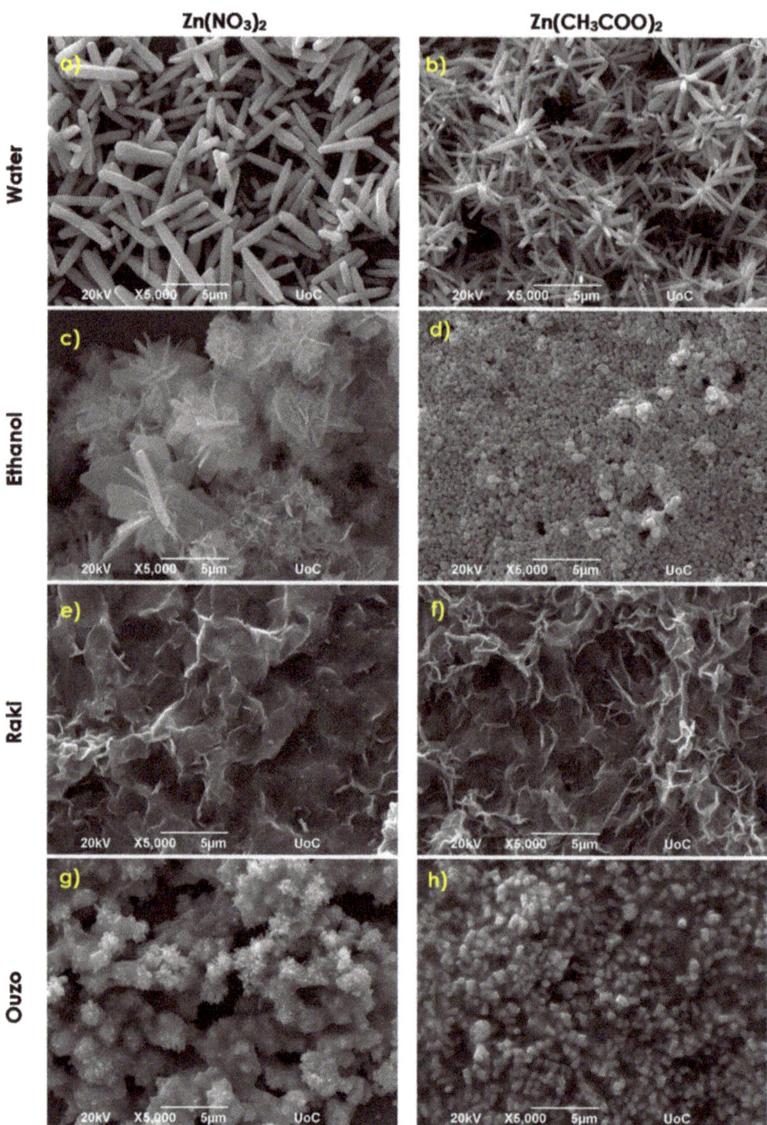

Figure 1. SEM images of ZnO samples synthesized via synthesis routes 1, 2 and 3 at 95 °C.

As it can be observed in Figure 1a, in a water solvent, the use of zinc nitrate and HMTA determined the growth of rod-like nanostructures with high aspect ratios for which the cross section is hexagonal and are randomly distributed onto the substrate. Thus, it seems that the obtained rods preserve the crystal structure of wurtzite ZnO, which is further highlighted from X-ray diffraction analysis. A careful inspection using the SEM analysis software provides an average length and width of ~5 μm and ~500 nm, respectively. Furthermore, the use of zinc acetate and HMTA resulted in flower-like structuring, as one can observe in Figure 1b. Flower-like structures are formed by hexagonal nanorods with a length of ~2–3 μm and a typical width of about 200–300 nm. In addition, several individual rods with ~2–3 μm mean length and larger width of about 300–400 nm can also be observed. It is worth noticing that the use of zinc acetate results in better substrate coverage of ~90%, while zinc nitrate use results in ~80% coverage. The use of ethanol as a solvent results in the formation of co-existing flakes and flower-like structures composed of flakes when using zinc nitrate, as presented in Figure 1c. Flakes thickness ranges between 200 and 800 nm, while their edge length is between 2 and 6 μm, and the substrate coverage is about 80%. By using zinc acetate, we obtained nanospheres with a mean diameter of ~500 nm that reaches a substrate coverage of around 98%. Using ethanol containing Raki traditional Greek drink as solvent results in the dense covering of the Si substrate with flake-like structures in both zinc source cases, as shown in Figure 1e,f. In the case of zinc acetate, the textures are crinklier, and the flakes are denser. Using Ouzo traditional Greek drink as solvent results in a completely different morphology, as presented in Figure 1g,h. For the use of zinc nitrate and HMTA, flower-like structures consisting of nanometer-scale hexagonal rods of typical length and width of ~1 μm and ~500 nm, respectively, were formed. The structuring changes to nanorods with a hexagonal cross section in presence of zinc acetate. The average length of nanorods is ~700–800 nm, while the average width is about 200 nm. The substrate coverage decreased from 80% to 60%, respectively. When the growth temperature becomes 195 °C, the ZnO structuring onto the Si substrate changes dramatically, as can be observed from Figure 2. Figure 2a shows tip rod-like ZnO microstructures resulting from the aqueous solution of zinc nitrate and HMTA. The tip rod length is ~1.5–3 μm, while the width is ~700 nm. Some bigger rod-like structures are also present, with a length of ~9 μm. In the case of zinc acetate such as a zinc source (Figure 2b), the structuring is flower-like, consisting of tip nanorods with an average length of ~3–5 μm and width of ~600–700 nm. A noticeable increase in substrate coverage from about 70% to 85% can be observed again for zinc acetate use. The use of ethanol results in structuring consisting of the coexistence of tip rods at a of length ~9 μm and width of ~2 μm and flower-like architectures consisting of tip rods with lengths and widths of about 7 μm and ~2 μm, respectively, for zinc nitrate, as shown in Figure 2c. When zinc nitrate is substituted by zinc acetate, the coating consists of spheres, with a mean diameter of ~500 nm (Figure 2d). With the exception of a completely different way of structuring, there is an increase in the coverage of the Si substrate from ~70% to 95%. In the case of Raki use as a solvent, there is a remarkable change in ZnO coating morphology. When zinc nitrate and HMTA are used at 195 °C, ZnO rod-like structures are grown on the Si(100), as observed in Figure 2e. These rods possess a hexagonal cross section, mean length of ~5 μm, and width of ~2 μm. Close morphology is observed when zinc acetate was used, but the rods become uncommon twin hexagonal-truncated-pyramid rod-like structures (Figure 2f). A similar morphology of ZnO was previously observed by Jun Zhang et al. and Suchea et al. [26,27]. The average length and width of the twin hexagonal-truncated-pyramid rod-like structures were ~4–5 μm and ~2 μm, respectively, while substrate coverage increased from ~94% to 98%. Using Ouzo as the solvent, the zinc nitrate-HMTA reaction resulted in spiked mace-like structures composed of rods with a mean length and width of ~2.5 μm and ~250 nm, respectively, as it is observed in Figure 2g. When zinc nitrate was replaced by zinc acetate, nanorods with hexagonal cross section were formed, possessing a mean length of ~600 nm and a width of ~200–300 nm (Figure 2h). The substrates coverage increased again from ~80% to ~85%.

Figure 2. SEM images of ZnO samples synthesized via synthesis routes 1, 2, and 3 at 195 °C.

Replacing HMTA with the nontoxic baking ammonia for 1, 2, and 3 synthesis conditions result in a completely different surface morphology of ZnO coatings. SEM characterization images of typical samples obtained at 95 °C are presented in Figure 3.

Figure 3a,c show that replacing HMTA with the nontoxic Baking ammonia for 1 and 2 synthesis conditions result in an obvious decrease in crystallinity and the formation of agglomerated flakes in both cases of using an aqueous solution of zinc nitrate and zinc acetate. The flakes have average widths and edge lengths of about 60 nm and 400 nm, respectively, for zinc nitrate and a mean width of ~100 nm and edge length ~600 nm for zinc acetate use (Figure 3b). The substrate coverage in both cases is ~60–70%. Using ethanol as the solvent resulted in the formation of larger flakes than in the water case with an average width ranging between ~150 and 400 nm and edge length of ~1–2 µm for zinc nitrate, as depicted in Figure 3c, and average width and edge length of about 100 nm and 600 nm, respectively, for zinc acetate use (Figure 3d). In both cases, substrate coverage was ~70–80%. Using Raki as the solvent, the zinc nitrate–baking ammonia reaction resulted in rough surfaces consisting of small agglomerations of about 2 µm diameter consisting of ZnO nanoparticles. When zinc nitrate was replaced by zinc acetate, the morphology remains the same, but the average diameter was ~1–2 µm. The substrates coverages remain ~70%

(Figure 3e,f). Using Ouzo as a solvent, the zinc nitrate-baking ammonia reaction resulted in spherical ZnO nanoparticles with a mean diameter size of ~2 μm (Figure 3g), while the zinc acetate–baking ammonia reaction resulted in a more compact coating comprising agglomerated flake structures (Figure 3h). In both cases, the substrates coverages range at ~70–80%.

Performing the same synthesis at a temperature of 195 °C results in the improvement of Si coating crystallinity. SEM images of similar growth conditions as above at 195 °C are presented in Figure 4.

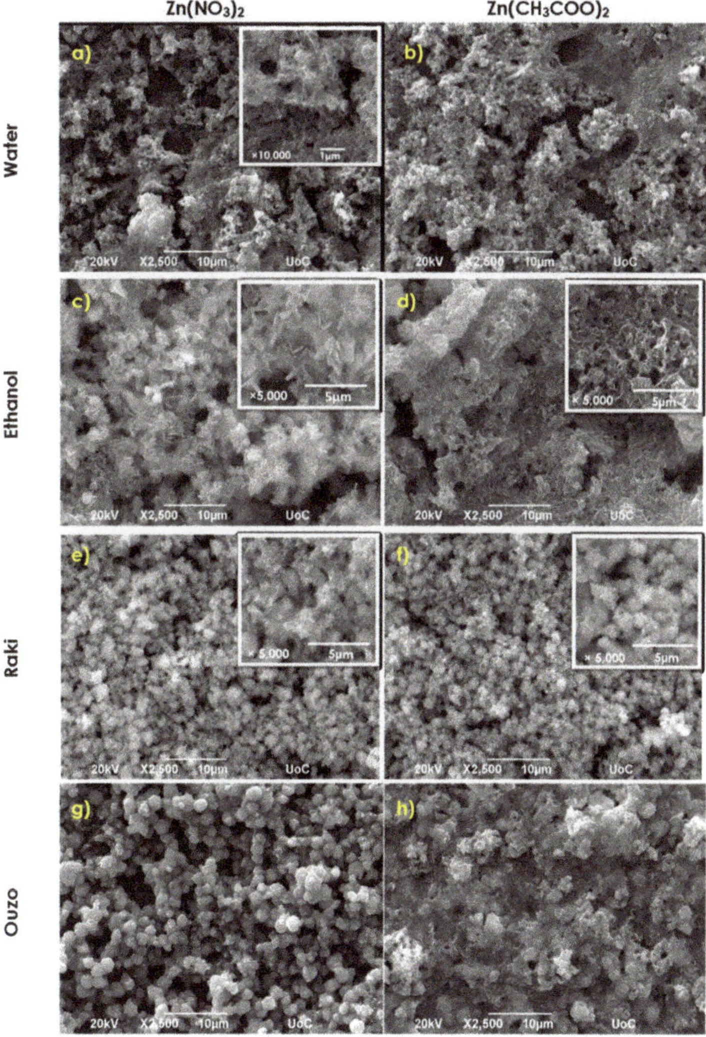

Figure 3. SEM images ×2500 of ZnO samples synthesized via synthesis route 4 at 95 °C. For the case of ZnO nanostructuring zoom images are inserted.

When water was used as a solvent, the presence of zinc nitrate–baking ammonia resulted in quite compact and rough coatings consisting of agglomerations of flakes (Figure 4a). The average width and edge length of flakes are ~100 μm and ~400 nm, respectively. The use of a solution of zinc acetate–baking ammonia in water at higher temperatures resulted in flower-like architectures (Figure 4b) consisting of tip nanorods with a length of several microns (~2–3 μm) and a typical width of around 500 nm. Isolated tip rods with a mean length of ~5–6 μm and width of about 400 nm can also be observed. With the exception of different morphologies, the substrate coverage decreases to 60% in the case of zinc acetate use. By using zinc nitrate and baking ammonia in ethanol, the coatings consist of nonhomogeneous ZnO tip rod flower-like structures (Figure 4c). Some of these structures are composed of dense rods and have a diameter ranging between 2 and 3 μm, while some others consist of more spare rods having a length of ~1 μm and a width of about 300 nm. When zinc nitrate was replaced by zinc acetate, ZnO short rods with hexagonal cross sections were formed (Figure 4d). The average length and width are 1–2 μm and ~300 nm–1 μm, respectively. The substrate coverage remained stable at ~70%. Figure 4e,f, depict the influence of Raki use on the morphology of the samples. More or less spherical agglomerations with rough surfaces and a mean diameter of ~2 μm were observed in both zinc source cases. The Si substrate coverage was ~75–80%. Zinc nitrate–baking ammonia in Ouzo reaction results in typical 1–2 μm diameter sea urchin-like ZnO spherical nanostructures that were formed (Figure 4g). In presence of zinc acetate, agglomerated flakes with typical width and edge length of ~100 nm and 1 μm, respectively, are formed (Figure 4h), resulting in an amorphous dense coating. To summarize, the increased reaction temperatures determine the improvement of coatings morphologies and structure evolution in almost all cases presented above.

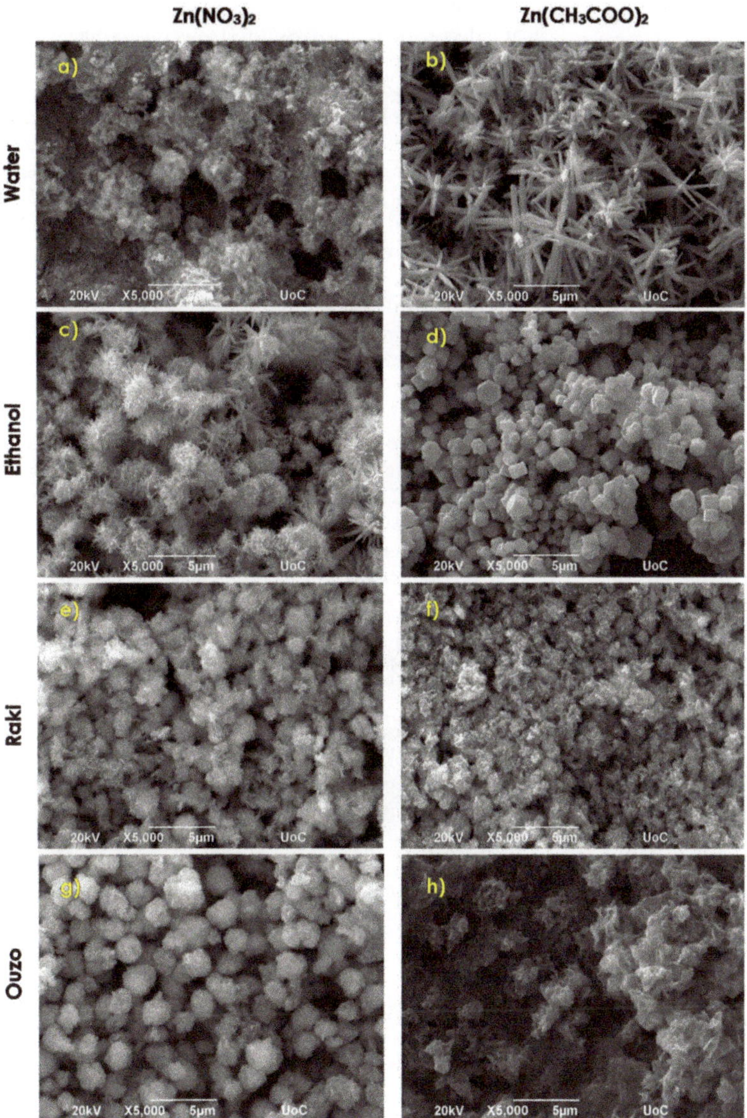

Figure 4. SEM images of ZnO samples synthesized via synthesis route 4 at 195 °C.

Using metallic Zn as a precursor in reaction conditions described in syntheses 5 and 6 determines a complete change of the ZnO growth onto the Si substrates. Changing the HMTA amount from 0.05 g to 0.15 g in synthesis 7 conditions also results in significant growth modifications. Typical SEM characterization images for materials obtained via these syntheses are presented in Figure 5.

Zinc powder with 0.05 g of HMTA dissolved in water resulted in sunflower-like ZnO structures with a diameter ranging between 2 and 5 μm (Figure 5a). As it is shown in Figure 5b, the increase in the amount of HTMA (0.15 g) did not affect the geometry of the structures, while their diameter increased up to ~7 μm. One noticeable difference is that when 0.05 g of HMTA was used, the sunflower-like structures were not well developed and, as a result, their size varies. The increased HMTA amount helps the homogenous growth of structures. The effect of HMTA quantity increase relative to substrate coverage is that it decreases from ~70–80% to a lower percentage. Figure 5c,d illustrate the influence of ethanol in the formation of ZnO structures. Mushroom-like architectures were developed with both amounts of HMTA. Similar morphologies have been previously reported by Jinzhou Yin et al. [28]. Substrate coverage increases from 40 to 60%, with an increasing the amount of HMTA. Using soda water as the solvent results in the formation of tip rods with a mean length of ~8 μm and width ~1 μm, as depicted in Figure 5e. Simultaneously with the presence of tip rods, agglomerated rods with hexagonal cross sections were also formed, as observed from the inset of the figure. The 0.15 g of HMTA determines the formation of hexagonal cross section rod-like structures, and the ~5 μm diameter flower-like agglomerations are formed by the flakes, as observed from Figure 5f. The rods forming the rod-like structures possess lengths of ~6 μm and widths of ~700 nm. The substrate coverage is about the same (80–85%). The use of lemon beverage results in flake structures for both amounts of HMTA (Figure 5g,h). The 0.05 g of HMTA resulting coating flakes show a width of~100 nm and an edge length of ~500 nm, while the respective dimensions when 0.15 g of HMTA was used are ~100 nm and ~600 nm. Finally, hydrogen peroxide (2.8% *w/w*) use determines the growth of coatings consisting of a combination of flakes and agglomerated hexagonal rods when 0.05 g of HMTA was used (Figure 5i). The width and edge length of flakes are ~250 nm and ~2 μm, respectively, while the rods have a length of ~1–2 μm and a width of ~800 nm. An increase in the amount of HMTA (0.15 g) resulted in coatings structured as clusters of rods, shown in Figure 5j, with an average diameter of ~5 μm and tip rods with a length of ~6 μm and width of ~2 μm.

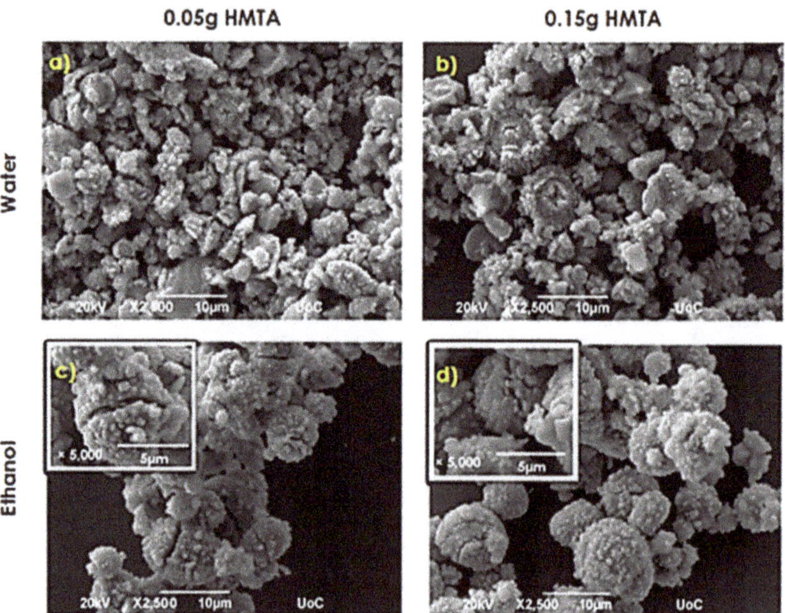

Figure 5. *Cont.*

Figure 5. SEM images ×2500 of ZnO samples synthesized via synthesis routes 6 and 7 for samples with smaller features zoom images are inserted.

By replacing the HMTA with 0.05 g and 0.15 g nontoxic baking ammonia in synthesis 5 and 6 conditions, the eco-friendly chemical synthesis effects on ZnO structuring onto Si substrates were studied. Figure 6 shows typical SEM images of the obtained coatings.

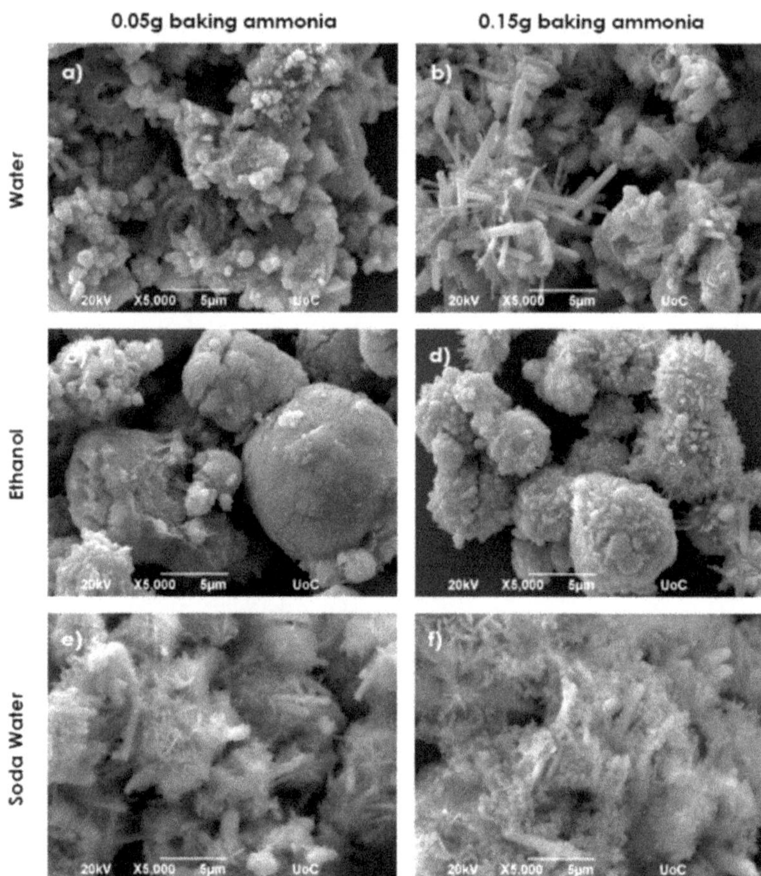

Figure 6. *Cont.*

Figure 6. SEM images of ZnO samples synthesized via synthesis route 8.

Using an aqueous solution of Zn metallic powder and baking ammonia, sunflower-like ZnO structures with a mean diameter of ~3–5 µm were formed (Figure 6a). The addition of more baking ammonia controls the formation of a mixture of various sized nanorods and irregular flower-like agglomerations, suggesting a change of growth mechanism, as observed in Figure 6b. Nanorods with a hexagonal cross sections possessing a length of ~8 µm and width of ~800–900 nm coexist with flower-like structures composed of flakes possessing a mean diameter of ~5 µm. Figure 6c presents the mushroom-like ZnO structures formed by the reaction of zinc powder with 0.05 g of baking ammonia in ethanol. Smaller mace-like structures with rough surfaces result when the amount of baking ammonia was increased (Figure 6d). The substrate coverage decreased as the amount of baking ammonia increased. Using soda water as the solvent resulted in the formation of rods, with an average length ~4 µm and width of ~600 nm, mixed in a mass of a structure composed of flakes with mean width and edge length of ~100 nm and ~600 nm, respectively (Figure 6e). Figure 6f illustrates the result of increasing the amount of baking ammonia; the rod-like structures with hexagonal cross sections developed while the background material appears to be a mixture of nanoparticles and flakes agglomerations. The substrate coverage increased (~90%). By using the lemon beverage as a solvent, fluffy coating structures consisting of thin flakes were formed for both baking ammonia amounts. The substrate coverage increased from 50% to 80% (Figure 6g,h). Irregular structures consisting of hexagonal rods randomly grown from a ZnO compact agglomeration background deposited onto the substrate were the result of hydrogen peroxide used as a solvent, as shown in Figure 6i,j.

Finally, baking ammonia was eliminated, and syntheses were performed by using the eco-friendly route 9. SEM images at ×2500 and ×5000 magnifications of the ZnO samples, prepared with only Zn metallic powder in various solvents via synthesis 9, are presented in Figure 7.

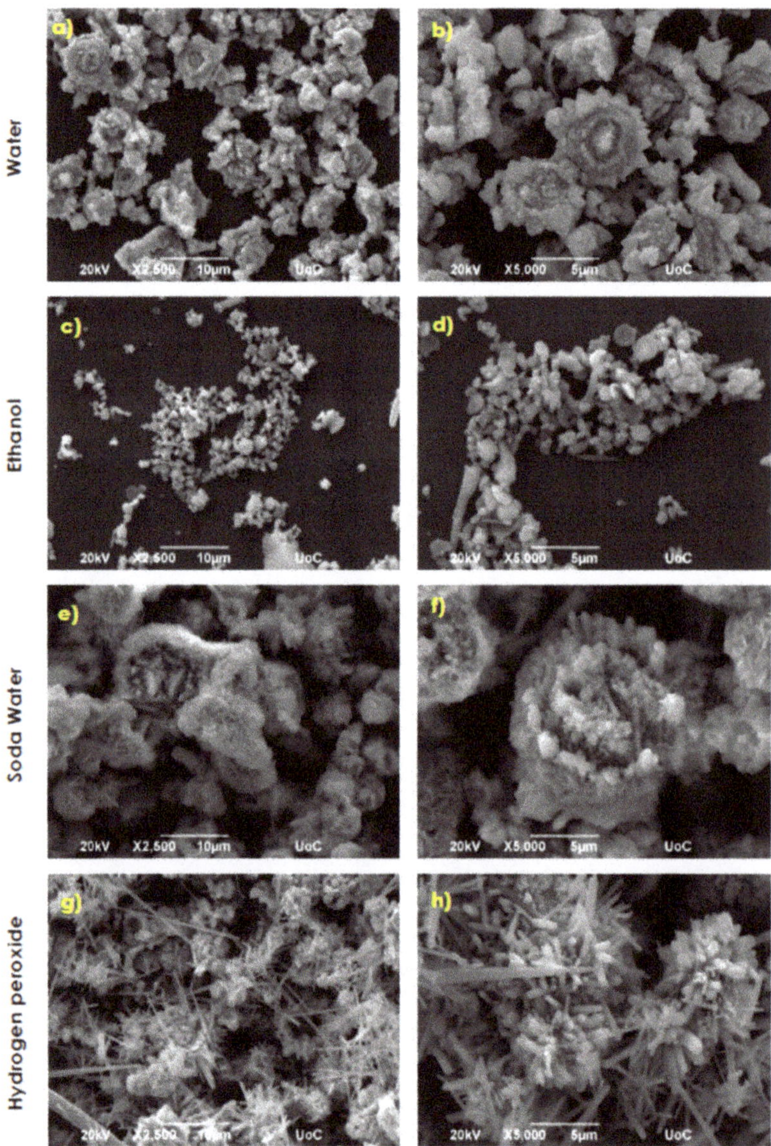

Figure 7. SEM images of ZnO samples synthesized via synthesis route 9 at 195 °C (**a**,**b**) two different magnifications for samples grown in water, (**c**,**d**) two different magnifications for samples grown in ethanol, (**e**,**f**) two different magnifications for samples grown in soda water (**g**,**h**) two different magnifications for samples grown in hydrogen peroxide.

Sunflower-like structures with irregular shapes of different sizes (Figure 7a,b) resulted from the use of aqueous solution as the solvent. The average diameter of these structures is ~6.5 μm. The substrate coverage can be estimated at about 40–50%. Figure 7c,d reveal only a very small coverage of the substrate area (~30–35%) with inhomogeneous ZnO nanoparticles. Small and larger rods, as well as hexagonal flakes, are formed. Soda water solvent use results in the growth of large flower-like structures with a mean diameter of ~10 μm. The flower-like structures consist of flakes with an average width of ~250 nm and an edge length of ~2 μm (Figure 7e,f). The substrate coverage reaches ~85–90%. The use of hydrogen peroxide (2.8% w/w) solvent results in coatings consisting of an inhomogeneous mixture of long tip rod structures mixed with spherical agglomerations of densely packed ZnO nanorods, as observed in Figure 7g,h. The tip rods have typical lengths of ~8 μm and a width of 500 nm, while the sphere-like structures are composed of nanorods with a mean length of ~1–2 μm and a width ~500 nm, respectively.

At this point, SEM characterization and micrograph analysis depict the profound relationship between ZnO morphology and the synthesis method, revealing an evolution from rod-like nanostructures to nanospheres or sunflower-like morphology. A further understanding of the microstructure can be attained by using X-ray diffraction, which can provide us with information regarding the interplanar distance, size of the crystalline domains, and the lattice strain of ZnO structuring onto Si substrates, as the commercially available high purity reagents used in typical ACG synthesis are replaced with cheaper nontoxic common ingredients coming from the local market in a trial confirming if nanostructured ZnO coatings suitable for optical and optoelectronic applications can be obtained in a cheap and eco-friendly nontoxic manner. Up to this point, from this study, one can observe that some of the proposed eco-friendly chemical routes are quite promising and further tuning of the reaction conditions may result in high quality nanostructured ZnO coatings onto crystalline Si substrates.

In addition to SEM studies, qualitative composition evaluation of all coatings was performed by EDX analysis. Due to the fact that most of the samples required preparation by Au metallization in order to avoid charging under electron beam exposure, EDX accurate quantitative estimation of elemental composition was not possible. Qualitative EDX analysis and rough stoichiometry estimation proved the presence of ZnO formation in all the fabricated samples. Coatings consisting of well-structured materials showed no presence of any other element except Si, Zn, and O. Coatings consisting of mixed morphology materials obtained via eco-friendly routes also showed the presence of some C contamination in various very low percentages. Some typical EDX spectra are presented in Figure 8.

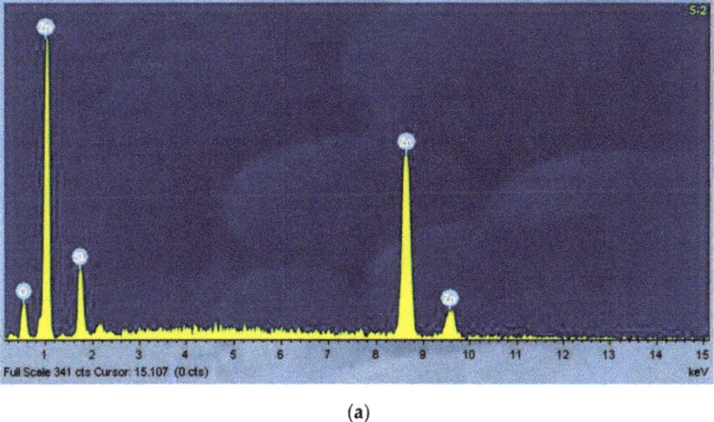

(**a**)

Figure 8. *Cont.*

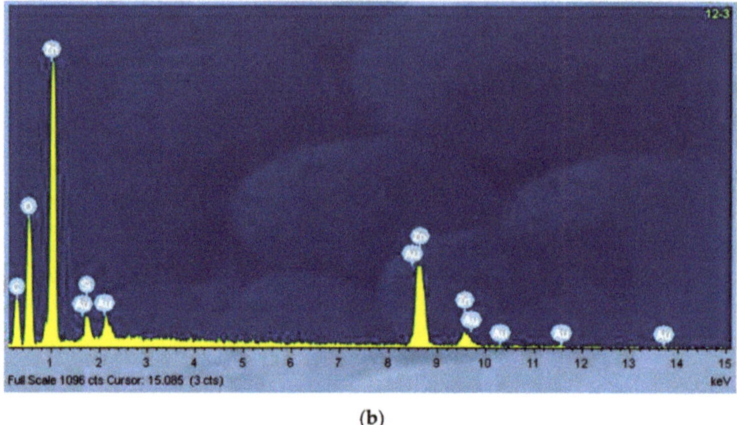

(b)

Figure 8. Typical EDX spectra of (**a**) ZnO samples synthesized with zinc nitrate and HMTA at 195 °C in ethanol solvent on Si(100) substrate and (**b**) ZnO samples synthesized with zinc acetate and HMTA at 95 °C in Ouzo solvent on Si(100) substrate.

3.2. XRD Characterization

XRD characterization of all synthesized materials was performed.

Materials resulting from synthesis 1 and 2 routes resulted in the formation of ZnO wurtzite phase according to JCPDS card no. 36-1451. In Figure 9a, the X-ray diffraction pattern presents a typical set of diffraction peaks of wurtzite ZnO (card no. 079-0208) that can be assigned to (100), (002), (101), (102), (110), (103), (200), (112), and (201) reflections, as observed in Figure 9. In the cases of synthesis 3 and 4 with the use of ethanol, Raki, and Ouzo as solvents, the presence of $Zn(OH)_2$ phase was observed, suggesting that ZnO syntheses were not complete. A slight improvement in crystallinity was observed for syntheses at higher temperatures in all situations. Two typical XRD spectra corresponding to the pure phase ZnO coating obtained from synthesis 2 (a) and to the incompletely formed ZnO phase coating obtained via synthesis 3 in the Ouzo solvent (b) are presented in Figure 9.

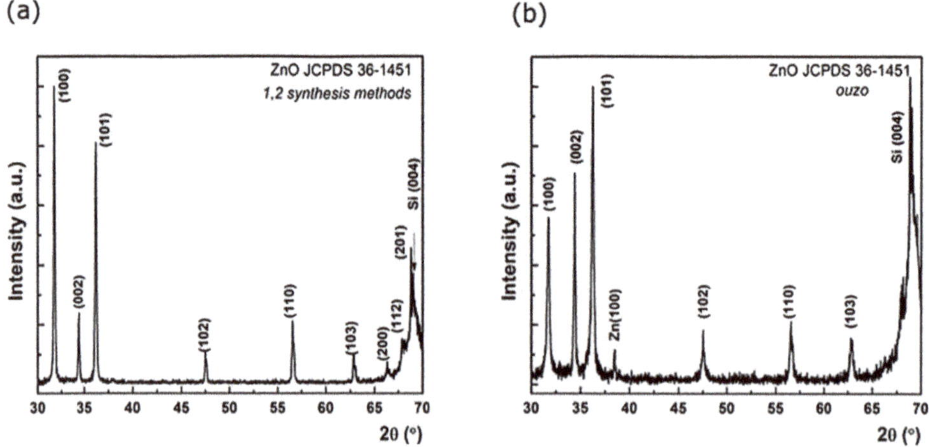

Figure 9. XRD spectra corresponding to the pure phase ZnO coating obtained from synthesis 2 (**a**) and to the incomplete formed ZnO phase coating obtained via synthesis 3 in Ouzo solvent (**b**).

Furthermore, once the salt precursors were replaced with metallic Zn powder precursor, the nanostructured ZnO coatings also show the occurrence of pure metallic Zn phases in certain situations. Figure 10 reveals XRD patterns for the samples obtained from zinc powder in different solvents.

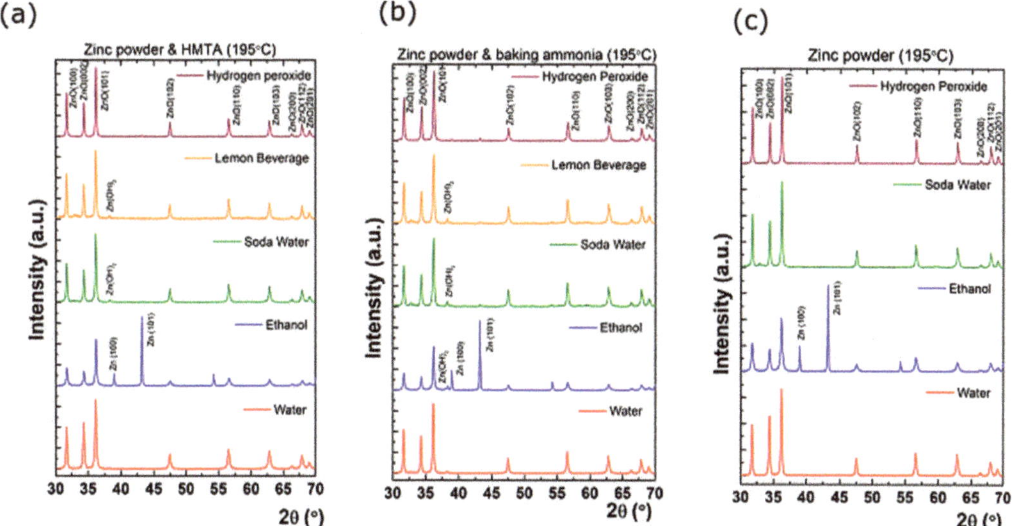

Figure 10. Grazing incidence X-ray diffraction for the synthesized samples from (**a**) Zn powder and Hexamethylenetetramine (HMTA), (**b**) Zn powder and baking ammonia, and (**c**) Zn powder at 195 °C in water, ethanol, soda water, lemon beverage, and hydrogen peroxide, respectively.

First, for the Zn powder and HMTA method, five solvents were used, namely water, ethanol, soda water, lemon beverage, and hydrogen peroxide. In the case of water, all diffraction peaks can be assigned unambiguously as ZnO. On the other hand, in the case of ethanol, unidentified diffraction peaks located at 38.99° and 43.19° can be observed. According to card no. 001-1238, these are consistent with metallic Zn with $a = b = 0.26$ nm and $c = 0.49$ nm. Furthermore, in the case of soda water and lemon beverage, only a small diffraction feature located at 38.30° corresponding to $Zn(OH)_2$ was detected. Finally, in the hydrogen peroxide solvent, pure ZnO was obtained. For the Zn powder and baking ammonia method, pure ZnO can be observed for water and hydrogen peroxide solvent. On the other side, in soda water and lemon beverage solvent, unreacted $Zn(OH)_2$ is observed, while a phase combination between metallic Zn and $Zn(OH)_2$ occurred in ethanol. For the last synthesis method, pure ZnO was obtained when water, soda water, and hydrogen peroxide solvents were used. At the same time, the presence of metallic Zn is observed again for ethanol. These results suggest the existence of three stages of ZnO formation: (i) full reaction of Zn(OH) in ZnO; (ii) combination between unreacted metallic Zn/ZnO; (iii) unreacted $Zn(OH)_2$/ZnO. For instance, a full reaction took place in the case of using water and hydrogen peroxide as solvents for Zn powder and HMTA and Zn powder and baking ammonia reactions, respectively. One can observe that the use of Zn powder also resulted in pure ZnO in the soda water solvent. In stage (ii), a combination between unreacted metallic Zn/ZnO occurred only for the ethanol solvent. Finally, a combination between unreacted $Zn(OH)_2$/ZnO appeared in stage (iii), and it is related to soda water and lemon beverage solvents for Zn powder and HMTA and Zn powder and baking ammonia use. The dependence of crystalline features and synthesis method, as well as the used solvent, is obvious. For a deeper understanding of the obtained ZnO crystallinity state and evolution, the size-strain plot Williamson–Hall method [29] was employed. This plot provides insight into how the residual strain in the crystallite distorts the crystal lattice and, hence, results in the broadening of diffraction peaks. In such cases, Williamson–Hall analysis should be used with Equation (1) by adding the two widths, given by crystallite size, τ and the lattice strain ε:

$$\beta cos\theta = \frac{k\lambda}{\tau} + 4\varepsilon sin\theta \tag{1}$$

where k is a shape factor of the crystallites taken as 0.9, and λ is the X-ray wavelength (e.g., 1.54 Å). By examining Equation (1), it is clear that the intercept of the linear fit is related to the mean crystallite size, while the slope determines the strain term. The XRD analyses associated Williamson–Hall plots are shown in Figure S1 in Supplementary Materials for each synthesis method as a red line. The values of the intercept and slope, as well as the fitting parameter R^2 that shows the goodness of fit, are listed in each case. Furthermore, the mean crystallite size and the lattice strain were determined and are tabulated in Table 1.

Table 1. Mean crystallite size and the lattice strain derived from Williamson–Hall plot.

Synthesis Method/Solvent	Mean Crystallite Size, d (nm)	Lattice Strain, ε (%)
Zn powder and HMTA/water	37.4	0.18
Zn powder and HMTA/ethanol	33.6	0.23
Zn powder and HMTA/soda water	40.7	0.11
Zn powder and HMTA/lemon beverage	40.7	0.13
Zn powder and HMTA/hydrogen peroxide	49.5	0.17
Zn powder and baking ammonia/water	42	0.12
Zn powder and baking ammonia/ethanol	33.8	0.09
Zn powder and baking ammonia/soda water	37.4	0.11
Zn powder and baking ammonia/lemon beverage	36.3	0.11
Zn powder and baking ammonia/hydrogen peroxide	43.3	0.16
Zn powder/water	39.6	0.17
Zn powder/ethanol	29.4	0.15
Zn powder/soda water	40.7	0.14
Zn powder/hydrogen peroxide	44.7	0.12

One can observe that the size of the crystalline domain spans over a wide range of values, from 29.4 nm up to 49.5 nm. At the same time, SEM analysis revealed complex morphologies: The smallest value of the crystalline domains corresponds to flower-like structures, while the largest one corresponds to a combination of flakes and agglomerated hexagonal rods. It is worth mentioning that the ethanol solvent resulted in the smallest mean crystallite size in the case of each synthesis method, which is ascribed to a poorer crystal quality in this case. At the same time, grazing incidence XRD patterns indicated the presence of metallic Zn for this case, ascribed above to stage (ii). On the opposite side, the best crystalline quality given by the use of hydrogen peroxide as a solvent is related to a full reaction of Zn(OH) to ZnO, as XRD analysis shows the presence of pure ZnO onto the substrate. Although the mean crystallite size has a large variation span from one synthesis to another, the lattice strain remains small for all materials (e.g., 0.1–0.2%). In the following, it will be shown that the different ZnO crystalline quality (i.e., different density of structural defects) would determine strong differences in photoluminescence (PL) spectra.

3.3. Photoluminescence Studies

As demonstrated above from scanning electron microscopy micrographs and X-ray diffraction analysis via Williamson–Hall plots, a variety of ZnO onto Si substrates sample morphologies was achieved. Such ZnO nanostructures have attracted much attention due to excellent high crystalline quality (mean crystallite size larger than 29 nm in each case, reaching even to 49 nm) and small lattice strain, which could be promising candidates for optical and optoelectronic applications.

The direct valence band of the wurtzite ZnO structure is split into three states, commonly called A (also called the light hole band), B (also called the heavy hole band), and C (also called the crystal-field split band) sub-bands, due to crystal field and spin–orbit interactions [2]. The luminescence properties of the obtained ZnO nanostructures have been intensively studied using photoluminescence spectroscopy. Generally, the typical photoluminescence spectrum of the nanostructured ZnO shows its band edge and exciton luminescence in the UV region. It also presents a green-centered broadband (GL) commonly related to deep-level defects [30]. The first of these bands is usually reported in the literature as NBE (near-band edge excitonic), while the second is known as the DLE band (deep-level emission) [31]. The sizeable binding exciton energy of ZnO (60 meV, nearly three times that of gallium nitride [30]) is responsible for the room temperature luminescence of this material, and it persists at temperatures as high as 700 K according to the literature [32]. Usually, the ratio between the integrated spectral intensity of NBE and DLE bands is used to estimate the contribution of recombination due to defect levels [33]. In ZnO nanostructures, the interaction between these bands becomes more complex because a high surface-volume ratio increases the density of surface defects, which strongly affects the processes of photoluminescence emission.

ZnO UV emission (also called near-band emission, NBE) is located close to its absorption edge (3.37 eV) and is produced by excitonic or band–band recombination. Bound excitons are extrinsic transitions and are related to dopants, native defects, or complexes, which usually create discrete electronic states in the bandgap [34] and, at low temperatures, are the dominant radiative channels. When the ZnO structure has high crystallinity, the UV emission is more intense than the emission in the visible region [35]. Although it is known that the NBE emission peak is composed of several peaks and shoulders of less intensity, in many works, its study is carried out indistinctly, analyzing only the sum of these emissions.

The photoluminescence (PL) of ZnO nanostructures also exhibits a DLE band due to the defect emissions. Several photoluminescence emission centers in the visible region are dependent on the synthesis technique and, hence, on the vacancies, surface defects, and morphology of the ZnO nanostructures. The ZnO visible PL mechanism is still far from being fully understood. The emissions in the region from 3.1 eV down to 1.653 eV are usually referred to as DLE. The deep levels denote the allowed levels inside the bandgap of the semiconductor that produces transitions with energy in the visible range of the spectrum. The band broadness is assumed to come from a superposition of many different deep levels (yellow peak, green peak, and blue peak) that emit simultaneously. Some previous work attributed the green and orange luminescence to extrinsic impurities such as Mo, Cu, Li, or Fe [35]. Various research studies show that undoped ZnO also presents photoluminescence peaks in the visible region and is generally ascribed to the electronic transition from single ionized VO^+ centers to the valence band edge [36,37].

Typical PL spectra of samples synthesized via synthesis routes 1, 2, and 3 at 95 °C and 195 °C are presented in Figure 11.

From Figure 11, one can clearly observe that even at RT almost all the samples exhibit ZnO UV emission NBE located close to its absorption edge (3.37 eV) and is produced by excitonic or band–band recombination. Bound excitons are extrinsic transitions and are related to dopants, native defects, or complexes, which usually create discrete electronic states in the bandgap. The only sample that had no PL emission is the one synthesized in Raki, which behavior is quite expected based on ZnO structuring onto the Si substrate. At RT, the UV emission is centered at 388 nm for water, 375 nm for ethanol, and 372 nm for Ouzo, which corresponds to energies of ~3.196 eV, ~3.307 eV, and ~3.333 eV, respectively. A better estimation of the energy band gap of ZnO can be given at low temperature PL measurements. At 13 K, the NBE emission is centered at 374 nm for water and 366 nm for ethanol and Ouzo. These wavelengths correspond to ~3.316 eV and ~3.388 eV, respectively, and are closer to the value of ~3.37 eV that is estimated as the NBE emission corresponding to the bound exciton (D^0X) of ZnO. In both spectra, one can observe a broadband emission from 450 nm to 700 nm, and the photoluminescence (PL) of ZnO nanostructures also exhibits a weak DLE band emission due to the defect emissions related to defects or impurities deriving from the solvents. At RT, the sample prepared in the aqueous solution exhibited the highest PL intensity, which at a lower temperature (13 K) was almost double.

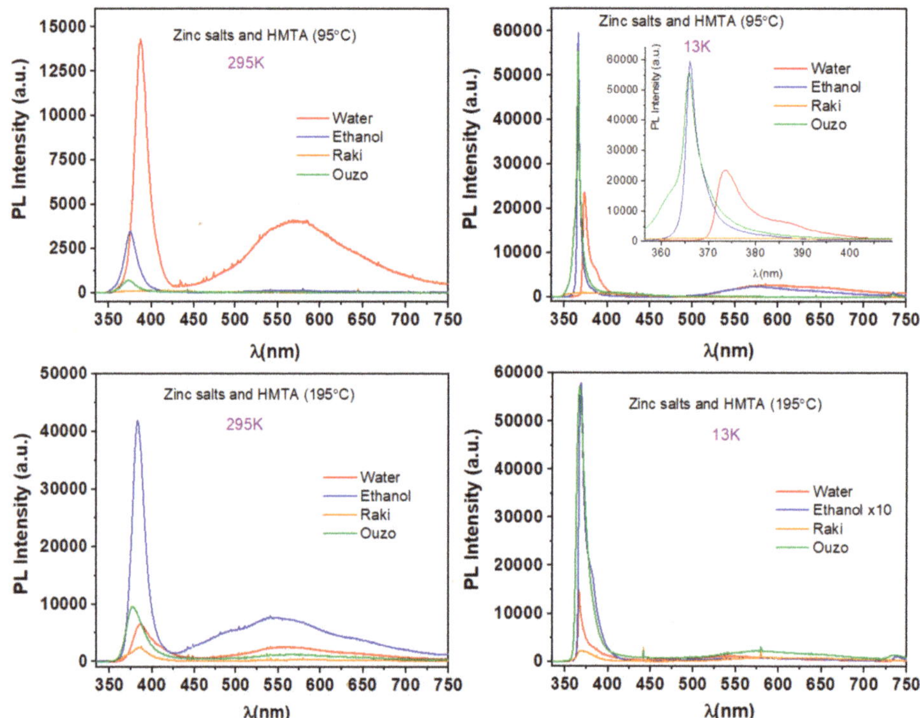

Figure 11. Typical PL spectra of samples synthesized via synthesis routes 1, 2, and 3 at 95 °C and 195 °C.

By increasing growth temperature, the ZnO samples' characteristic PL emission strongly changed for all kinds of solvents. At RT, the PL spectra included both the NBE, as well as the DLE (450 nm–650 nm) emissions; at 13 K, the visible emission considerably diminished. The UV emission at RT is centered at 387 nm for water and raki, 382 nm for ethanol, and 377 nm for Ouzo, and the corresponding energies are 3.204 eV, 3.246 eV, and 3.289 eV, respectively. At 13 K, the NBE peaks are centered at 367 nm for water and Ouzo and at 368 nm for ethanol and raki. The corresponding energies are 3.379 eV (water and Ouzo) and 3.370 eV (ethanol and raki), respectively. It is observed that ZnO synthesized in ethanol at 195 °C exhibits the strongest UV emission, which at 13 K becomes almost 14 times higher and, in the spectrum, an obvious shoulder becomes visible at ~380 nm wavelength near the central peak. This corresponds to ~3.26 eV. This value may be associated with a first-order longitudinal optical (LO) replica (DAP-LO). The shift of the two peaks is near the theoretical LO phonon energy in ZnO (70–75 meV) [38]. The broadening of the NBE peak as the temperature increases (see Figure 11) is attributed to the thermal decomposition and ionization of bound excitons, as well as to the strong coupling of phonons and excitons [38]. The photoluminescence intensity increases as the temperature is lowered to 13 K as a consequence of temperature quenching effects.

Replacing HMTA with nontoxic baking ammonia for 1, 2, and 3 synthesis conditions results in completely different surface morphology of ZnO coatings, and their PL characterization also reveals these changes. Typical PL spectra of samples synthesized via synthesis route 4 at 95 °C and 195 °C are presented in Figure 12.

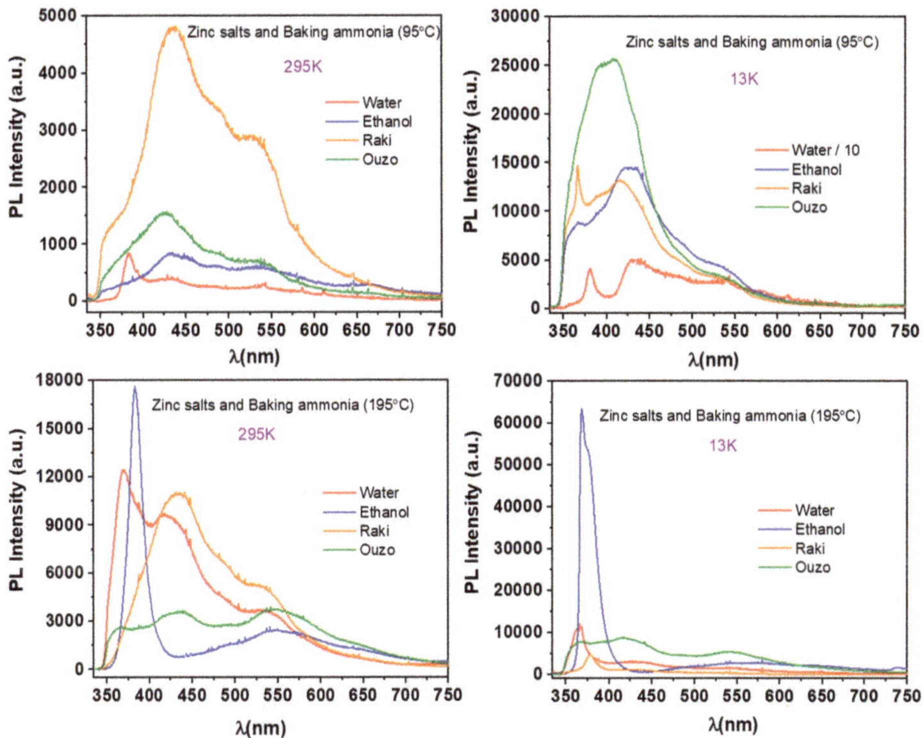

Figure 12. Typical PL spectra of samples synthesized via synthesis route 4 at 95 °C and 195 °C.

As it can be observed, in this case only the sample synthesized in aqueous solution exhibited low intensity NBE emission. The emission is centered at 382 nm, corresponding at 3.246 eV; its energy is close enough to the energy gap of ZnO. The samples prepared in ethanol, Raki, and Ouzo did not emit in the UV spectral region. Surprisingly, at 13 K, ZnO nanostructured materials synthesized in water and Raki solvents show clear NBE emission. The UV emission for water is centered at 381 nm and 366 nm for Raki, corresponding to energies of 3.255 eV and 3.388 eV, respectively. When the synthesis temperature was 195 °C, the PL performance of the samples improved, as observed in Figure 12. At RT, the ZnO sample synthesized in ethanol has a strong UV emission, accompanied by a weak visible emission. The UV emission is centered at 387 nm for water and 382 nm for ethanol, corresponding to 3.204 eV and 3.246 eV, respectively. At low temperature (13 K), the UV emission of ZnO samples synthesized in water is centered at 363 nm, 368 nm for ethanol, and 378 nm for Raki, corresponding to energies of 3.416 eV, 3.370 eV, and 3.280 eV, respectively. The best PL performance was exhibited to be synthesized in ethanol, and the emission intensity increased about four times when the PL temperature decreased at 13 K.

Using metallic Zn as a precursor in reaction conditions described in synthesis 5 and 6 determined a complete change of ZnO growth onto the Si substrates, and this reflects as well on the PL characteristic emission of the respective ZnO nanostructured coatings, as shown in Figure 13.

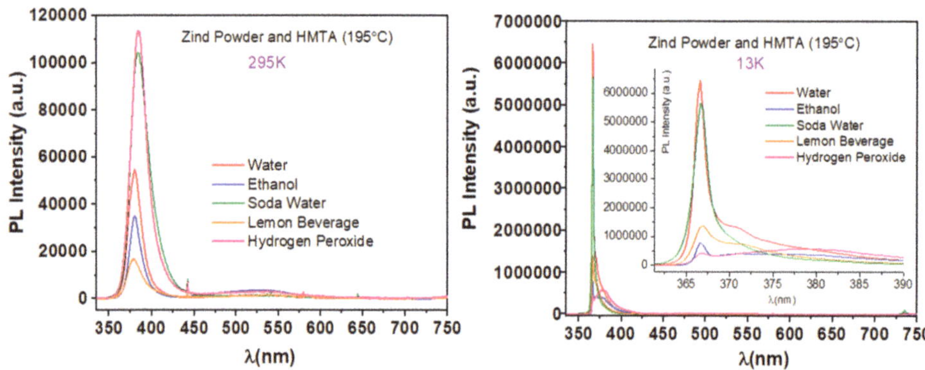

Figure 13. Typical PL spectra of samples synthesized via synthesis route 6 at 195 °C.

Even at RT, all the samples synthesized in different solvents exhibit clear the characteristic NBE emission of ZnO along with a very weak broad emission in the visible spectral region. At RT, the UV emission is centered at 380 nm for water and ethanol, at 384 nm for soda water, at 379 nm for the lemon beverage, and 385 nm for hydrogen peroxide (2.8% w/w), corresponding to the energies of 3.263 eV, 3.229 eV, 3.272 eV, and 3.221 eV, respectively. At 13 K, all of the peaks attributed to NBE emission are centered at 367 nm, which corresponds to 3.380 eV. The highest PL intensity, at RT, is exhibited by the sample prepared with hydrogen peroxide (2.8% w/w), which increased four times at 13 K.

The effects of replacing the HMTA with nontoxic baking ammonia in synthesis 5 and 6 conditions and the eco-friendly chemically synthesized ZnO onto Si substrates PL emission spectra were studied and are presented in Figure 14.

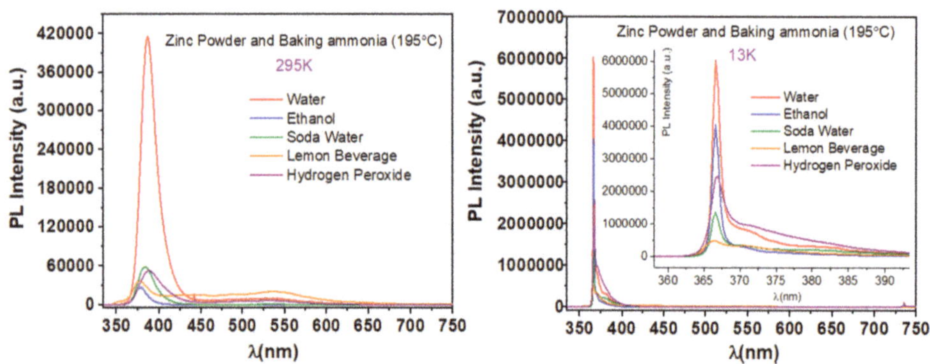

Figure 14. Typical PL spectra of samples synthesized via synthesis route 8 at 195 °C.

One can easily notice from Figure 14 that all of the prepared samples, independently of the solvent, have the characteristic UV emission of ZnO at both temperatures. Furthermore, at RT, a very weak broad emission at the visible spectral region is observed. At RT, the UV emission is centered at 387 nm for water and hydrogen peroxide (2.8% w/w), at 379 nm for ethanol, at 384 nm for soda water, and 378 nm for lemon beverage, corresponding to the energies of 3.204 eV, 3.272 eV, 3.229 eV, and 3.280 eV, respectively. At 13 K the characteristic NBE emission peaks are centered at 367 nm for water, ethanol, soda water, and hydrogen peroxide (2.8% w/w) and 366 nm for the lemon beverage. The corresponding energies are 3.380 eV and 3.390 eV. The highest PL emission intensity at RT can be observed for the sample prepared with water, and its intensity increases 15 times at 13 K.

Finally, after the baking ammonia was eliminated, and syntheses were performed using the eco-friendly route 9; the obtained nanostructured ZnO materials show the PL characteristic emission spectra presented in Figure 15.

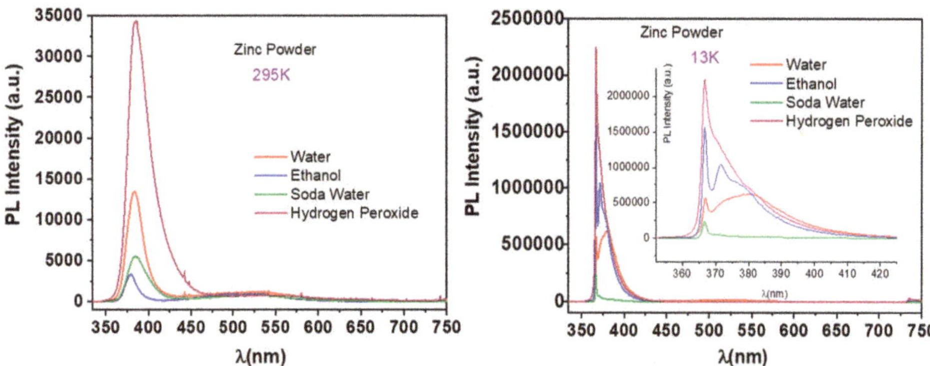

Figure 15. Typical PL spectra of samples synthesized via synthesis route 9 at 195 °C.

All the prepared samples, independent of the solvent used, show the characteristic UV emission of ZnO at both temperatures. At RT, the very weak broad emission in the visible spectral region is observed. At RT, UV emission is centered at 384 nm for water and soda water, at 380 nm for ethanol, and 386 nm for hydrogen peroxide (2.8% w/w), corresponding to the energies of 3.229 eV, 3.263 eV, and 3.212 eV, respectively. When the temperature decreased at 13 K, all the peaks attributed to exciton recombination are centered at 367 nm, which corresponds to 3.380 eV. The highest PL intensity at RT is shown by the sample prepared with hydrogen peroxide (2.8% w/w), and it increases 63 times at 13 K.

Using the PL results, an estimation of quantum yield (QY) can be obtained by evaluating the ratio of RT/LT PL intensity integration. This estimation results in the following observations:

(1)   The samples with Zinc salts, HMTA, or Baking soda (Figures 11 and 12) show a QY of 10–30%, the highest value for the sample prepared in ethanol.
(2)   The samples with Zinc powder, HMTA, or Baking soda (Figures 13 and 14) results in QY ranging from 4% up to 20%, with the best value for the sample prepared in hydrogen peroxide.
(3)   The samples synthesized via synthesis route 9 (Figure 15) show QY values ranging from 3% up to 20%, with the better value corresponding to the sample prepared in soda water.

Figure 16 summarizes, in a comparative manner, the previous results regarding the PL performance for all the samples synthesized during this study. Some samples exhibited a very strong PL characteristic NBE emission peak.

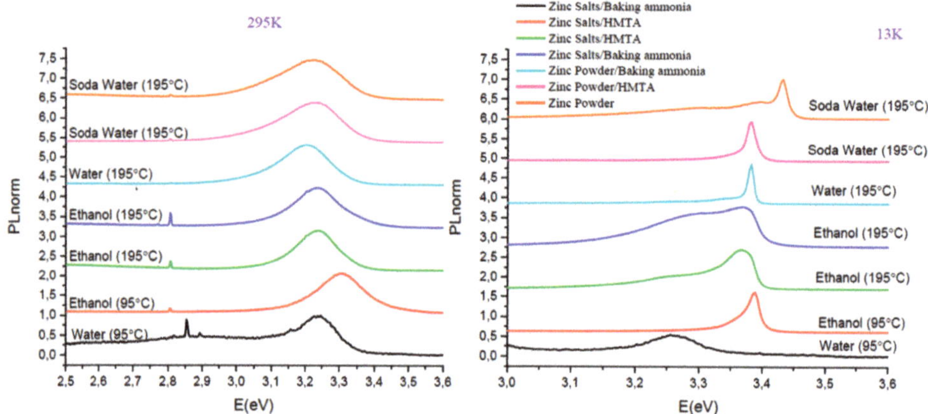

Figure 16. Summary of PL performance for all the samples synthesized during this study.

As it can be observed, NBE emission at RT position changes from 3.224 eV (light blue line: zinc powder/baking ammonia in water at 195 °C) to 3.303 eV (red line: zinc salts/HMTA in ethanol solution (95 °C)). The FWHM of the NBE emission PL peak provides information about the homogeneous/inhomogeneous broadening of the natural linewidth of the transition, which can be related to crystallinity disorder. First, validation of the line shape could be conducted by using a Voigt function with a different ratio of Lorentzian/Gaussian component. For instance, a Lorentzian is an ideal line shape, and deviation to a Gaussian indicated disorder. A PL spectrum may also have phonon side bands. The inhomogeneity might be caused by band gap fluctuations, which can be due to chemical fluctuations or electrostatic potential variations. The latter can also be related to the degree of compensation, i.e., the ratio of donors to acceptors. If there is a high density of donors or acceptors locally, which cannot be screened, then the band gap fluctuations are observed, resulting in the broadening of the NBE PL peak. Hence, the FWHM of a peak also provides insight into the doping of the investigated material. If the material is strained and/or elongated, it is known that a built-in (or external) (piezo) electric field results in tilting of the energy levels and, thus, in the broadening of PL peaks. The same effect appears whenever some charges are present around, for example, the surface. Moreover, the increase in laser power in the PL experiment can also result in the broadening of the NBE peak since more and more levels are filled; thus, as the samples are formed of many ZnO nanostructures, their collective emission will blend, and the peaks may become broader. The FWHM of the peaks was evaluated to be about 200 meV for all samples, and a Lorentzian fit for the NBE emission peak was used for all samples. From low temperature PL spectra (13 K), the strong emission observed at ~3.38 eV corresponds to free exciton (FX), as well as emission at lower energy which can be assigned to the donor-bound exciton (D-X) and phonon-assisted transitions. The sample synthesized in aqueous solutions of zinc salts and baking ammonia (black line) does not show the FX emission, while the sample prepared with zinc powder in soda water solvent shows a blue shift 50 meV from the FX peaks. The observation of the emission associated with FX, the high PL intensity, and the small FWHM indicates good crystal quality of ZnO nanostructures, even in the samples grown in solvents such as soda water.

The best PL performance was detected on the following.

ZnO synthesized by zinc salts and HMTA at 95 °C in ethanol solution (QY ~30%).

ZnO synthesized by zinc salts and baking ammonia at 195 °C in ethanol solution (QY ~30%).

ZnO synthesized by zinc powder at 195 °C in soda water solution (QY ~20%).

Finally, one can observe that for, the Zn powder and HMTA method, X-ray diffraction analysis showed the best crystal quality for the hydrogen peroxide solution, for which the mean crystallite size **d** is 49.5 nm; the crystal quality then decreased systematically when using soda water, lemon beverage, water, and ethanol. In the case of ethanol, the mean crystallite size reached 33.6 nm, which corresponds to a relative decrease of 32%. At the same time, using hydrogen peroxide and soda water as solvents also resulted in the highest PL intensity at 295 K. Moving forward to the Zn powder and Baking ammonia method, using the water as a solvent resulted in high crystal quality (e.g., **d** = 42 nm) and concomitantly the highest PL intensity. At the opposite side, ethanol solvent resulted in the worst crystal quality (e.g., **d** = 33.8 nm), as well as the smallest PL intensity. For the last synthesis method, namely the zinc powder method, ZnO with the best crystal quality (e.g., **d** = 44.7 nm) was obtained in hydrogen peroxide, and it is related to the highest PL intensity at 295 K. Moreover, ZnO with the worst crystal quality was obtained in ethanol, presenting also the smallest PL intensity. In light of the above information, it seems that the crystal quality that it is related to the PL intensity, together with the microstructure, plays a pivotal role in designing devices with targeted physical properties.

4. Conclusions

This study is focused on the development of various kinds of nanostructured ZnO onto Si substrates via chemical route synthesis using both classic solvents, such as raki and Ouzo, as well as of some usual non-toxic beverages, such as soda water, lemon beverage, or hydrogen peroxide. In this manner, the expensive high purity reagents acquired from specialized providers were successfully substituted in ACG synthesis of ZnO. Scanning electron microscopy micrographs reveal the close relationship between ZnO morphology and the synthesis method, revealing an evolution from rod-like nanostructures, to nanospheres or sunflower-like morphology. Since the microstructure determines different optical properties, disclosure of the main structural parameters becomes mandatory. In this sense, the Williamson–Hall method was used to provide a separate description of the effects given by the size and strain in the total broadening of the diffraction peaks. For instance, for the Zn powder and HMTA method, X-ray diffraction showed the best crystal quality for hydrogen peroxide solution for which the mean crystallite size **d** is 49.5 nm; then, the crystal quality decreased systematically when used soda water, lemon beverage, water, and ethanol. In the case of ethanol, the mean crystallite size reached 33.6 nm, which corresponds to a relative decrease of 32%. At the same time, using the hydrogen peroxide and soda water as solvents also resulted in the highest PL intensity at 295 K. Moving forward to the Zn powder and baking ammonia method, using water as solvent resulted in high crystal quality (e.g., **d** = 42 nm) and concomitantly the highest PL intensity. At the opposite side, ethanol solvent resulted in the worst crystal quality (e.g., **d** = 33.8 nm), as well as the smallest PL intensity. For the last synthesis method, namely zinc powder method, ZnO with the best crystal quality (e.g., **d** = 44.7 nm) was obtained in hydrogen peroxide, and it is related to the highest PL intensity at 295 K. Moreover, ZnO with the worst crystal quality was obtained in ethanol, presenting also the smallest PL intensity. Due to large surface-to-volume ratios and individual highly crystalline ZnO nanostructures, PL investigations proved that direct band-gap ZnO growth onto the Si substrates possesses wurtzite type structure, suitable for applications that require complex surface structuring with a high number of grain boundaries, such as nonlinear optics and some optoelectronic applications.

In summary, in the first set of experiments, zinc nitrate and zinc acetate (zinc salts) and HMTA were reacted in water, ethanol, Raki, and Ouzo solvents. Each synthesis was performed at 95 °C and 195 °C for 2 h. The morphology of the nanostructured ZnO onto Si coatings was characterized by diversity, depending on the zinc precursor and the solvent used, as well as the synthesis temperature. Higher temperatures resulted in the formation of ZnO nanostructures with better crystallinity. A typical example is a sample prepared using Raki as the solvent; the flake-like structure was converted to hexagonal section rods when the synthesis temperature increased from 95 °C to 195 °C. Regarding crystallinity, XRD patterns revealed the hexagonal wurtzite phase of ZnO. The PL characterization of the samples synthesized at 95 °C revealed higher emission intensity from the sample synthesized in aqueous solutions, while the increase in the synthesis temperature rendered ethanol solvent the best. This can be correlated to the ZnO nanostructuring onto the Si substrate and the crystalline structure. Furthermore, the HMTA was replaced by baking ammonia. The derived morphologies were different from the ones that resulted when HMTA was used instead of baking ammonia. The structural characterization of the synthesized samples revealed the existence of $Zn(OH)_2$ byproducts along with ZnO. The existence of these byproducts was confirmed by PL characterization. Broad and intense PL emission peaks existed on the visible region of the electromagnetic spectrum. Furthermore, according to the PL characterization, only water and ethanol were the solvents that resulted in materials with UV emission at RT. After the substitution of HMTA by baking ammonia, the zinc salts were substituted by zinc powder. SEM images proved that the use of different solvents resulted in different ZnO structuring, while the XRD patterns revealed the hexagonal wurtzite phase of ZnO. All the obtained samples exhibited the UV emission at RT according to the PL characterization, but soda water and hydrogen peroxide used as solvents permitted the growth of materials with the strongest NBE emission. The simple substitution of HMTA by baking ammonia and the use of zinc sources in the same solvents similar to before results in a new kind of ZnO structuring with new properties. Upon changing the solvent and the amount of baking ammonia, different ZnO morphologies were derived. The UV emission peak was present in the PL spectra of the samples at RT, with higher intensities for the nanostructured ZnO synthesized in water. The last set of studied materials included samples prepared by decomposition of zinc powder in water, ethanol, soda water, and hydrogen peroxide (2.8% w/w). The morphology changes again according to the different solvents, while XRD characterization revealed the hexagonal wurtzite phase of ZnO. All the samples were characterized by PL spectroscopy at room temperature and 13 K. Typical features, such as NBE and DLE bands, were observed, and different contributions of DLE intensity were interpreted as measures of structural defect densities; the UV emission corresponding to the NBE emission of ZnO was observed at RT, but the sample prepared in hydrogen peroxide solvent had the strongest emission. It was proved that a large variety of ZnO nanostructures and individual nanostructures with high crystallinity and excellent optical properties can be obtained by using some very cheap and facile chemical synthesis routes with high reliability. Further optimization of the desired synthesis route can tune the material properties to achieve the necessary quality needed for a specific application.

The wide range of ZnO morphologies induces different sizes of the crystalline domains, and we proved the further relationship relative to the optical properties. In this context, it is clear that the use of different synthesis methods leads to a tuning of main structural parameters that are further related to different optical properties. This approach can help design devices with targeted optical properties using simple chemical methods.

Supplementary Materials: The following are available online at https://www.mdpi.com/article/10.3390/nano11102490/s1, Figure S1: Wiliamson-Hall plot (red line) for the synthesized samples from Zn powder & Hexamethylenetetramine (HMTA)—column 1, Zn powder & baking ammonia—column 2 and Zn powder at 195 °C in water, ethanol, soda water, lemon beverage and hydrogen peroxide, respectively—column 3. The value of the intercept and slope, as well as the fitting parameter R^2, that shows the goodness of fit are listed in each case.

Author Contributions: Conceptualization, M.P.S. and G.K.; data curation, M.P.S., E.P., C.R., M.A., Z.V., R.I., and P.P.; formal analysis, M.P.S., E.P., C.R., M.A., A.M., Z.V., R.I. and P.P.; investigation, M.P.S., E.P., M.A., A.M., Z.V., R.I., P.P. and G.K.; methodology, G.K.; project administration, G.K.; resources, G.K.; supervision, M.P.S. and G.K.; validation, M.P.S. and G.K.; visualization, M.P.S. and A.M.; writing—original draft, M.P.S., C.R., P.P. and G.K.; writing—review and editing, M.P.S. and G.K. All authors have read and agreed to the published version of the manuscript.

Funding: This research received no specific external funding.

Institutional Review Board Statement: Not applicable.

Informed Consent Statement: Not applicable.

Data Availability Statement: The raw and processed data required to reproduce these findings cannot be shared at this time due to technical or time limitations. The raw and processed data will be provided upon reasonable request to anyone interested anytime, until technical problems are be solved.

Acknowledgments: Part of this work was co-financed by the European Union and Greek national funds through the Operational Program Competitiveness, Entrepreneurship and Innovation, under the call RESEARCH–CREATE–INNOVATE (acronym: POLYSHIELD; project code: T1EDK-02784). P. Pascariu contribution was partially supported by project number PN-III-P1-1.1-TE-2019-0594 of the Romanian Ministry of Research, Innovation and Digitization, CNCS/CCCDI—UEFISCDI, within PNCDI III. M. P. Suchea and C. Romanitan contributions were partially supported by the Romanian Ministry of Research, Innovation and Digitalisation thorough "MICRO-NANO-SIS PLUS" core Programme.

Conflicts of Interest: The authors declare that they have no known competing financial interest or personal relationships that could have appeared to influence the work reported in this paper.

References

1. Klingshirn, C.; Fallert, J.; Zhou, H.; Sartor, J.; Thiele, C.; Maier-Flaig, F.; Schneider, D.; Kalt, H. 65 years of ZnO research—Old and very recent results. *Phys. Status Solidi B* **2010**, *247*, 1424–1447. [CrossRef]
2. Özgür, U.; Alivov, Y.I.; Liu, C.; Teke, A.; Reshchikov, M.A.; Doğan, S.; Avrutin, V.; Cho, S.J.; Morkoç, H. A comprehensive review of ZnO materials and devices. *J. Appl. Phys.* **2005**, *98*, 041301. [CrossRef]
3. Shahzad, S.; Javed, S.; Usman, M. A Review on Synthesis and Optoelectronic Applications of Nanostructured ZnO. *Front. Mater.* **2021**, *8*, 613825. [CrossRef]
4. Rodrigues, J.; Fernandes, A.J.S.; Monteiro, T.; Costa, F.M. A review on the laser-assisted flow deposition method: Growth of ZnO micro and nanostructures. *CrystEngComm* **2019**, *21*, 1071–1090. [CrossRef]
5. Znaidi, L. Sol–gel-deposited ZnO thin films: A review. *Mater. Sci. Eng. B* **2010**, *174*, 18–30. [CrossRef]
6. Arya, S.; Mahajan, P.; Mahajan, S.; Khosla, A.; Datt, R.; Gupta, V.; Young, S.-J.; Oruganti, S.K. Review—Influence of Processing Parameters to Control Morphology and Optical Properties of Sol-Gel Synthesized ZnO Nanoparticles. *ECS J. Solid State Sci. Technol.* **2021**, *10*, 023002. [CrossRef]
7. Hasnidawani, J.; Azlina, H.; Norita, H.; Bonnia, N.; Ratim, S.; Ali, E. Synthesis of ZnO Nanostructures Using Sol-Gel Method. *Procedia Chem.* **2016**, *19*, 211–216. [CrossRef]
8. Harun, K.; Hussain, F.; Purwanto, A.; Sahraoui, B.; Zawadzka, A.; Mohamad, A.A. Sol–gel synthesized ZnO for optoelectronics applications: A characterization review. *Mater. Res. Express* **2017**, *4*, 122001. [CrossRef]
9. Nowak, E.; Szybowicz, M.; Stachowiak, A.; Koczorowski, W.; Schulz, D.; Paprocki, K.; Fabisiak, K.; Los, S. A comprehensive study of structural and optical properties of ZnO bulk crystals and polycrystalline flms grown by sol-gel method. *Appl. Phys. A* **2020**, *126*, 552. [CrossRef]
10. Marouf, S.; Beniaiche, A.; Guessas, H.; Azizi, A. Morphological, Structural and Optical Properties of ZnO Thin Films Deposited by Dip Coating Method. *Mater. Res.* **2017**, *20*, 88–95. [CrossRef]
11. Haque, M.J.; Bellah, M.M.; Hassan, M.R.; Rahman, S. Synthesis of ZnO nanoparticles by two different methods & comparison of their structural, antibacterial, photocatalytic and optical properties. *Nano Express* **2020**, *1*, 010007.
12. da Silva, E.C.; de Moraes, M.O.S.; Brito, W.R.; Passos, R.R.; Brambilla, R.F.; da Costa, L.P.; Pocrifka, L.A. Synthesis of ZnO nanoparticles by the sol-gel protein route: A viable and efficient method for photocatalytic degradation of methylene blue and ibuprofen. *J. Braz. Chem. Soc.* **2020**, *31*, 1648–1653. [CrossRef]
13. Darshita, M.N.; Sood, R. Review on synthesis and applications of zinc oxide nanoparticles. *Preprints* **2021**, 2021050688.
14. Pomastowski, P.; Król-Górniak, A.; Railean-Plugaru, V.; Buszewski, B. Zinc Oxide Nanocomposites—Extracellular Synthesis, Physicochemical Characterization and Antibacterial Potential. *Materials* **2020**, *13*, 4347. [CrossRef] [PubMed]
15. Kolodziejczak-Radzimska, A.; Jesionowski, T. Zinc Oxide—From Synthesis to Application: A Review. *Materials* **2014**, *7*, 2833–2881. [CrossRef]
16. Amakali, T.; Daniel, L.S.; Uahengo, V.; Dzade, N.Y.; de Leeuw, N.H. Structural and Optical Properties of ZnO Thin Films Prepared by Molecular Precursor and Sol–Gel Methods. *Crystals* **2020**, *10*, 132. [CrossRef]
17. Vernardou, D.; Kenanakis, G.; Couris, S.; Koudoumas, E.; Kymakis, E.; Katsarakis, N. pH effect on the morphology of ZnO nanostructures grown with aqueous chemical growth. *Thin Solid Films* **2007**, *515*, 8764–8767. [CrossRef]
18. Available online: https://www.accessdata.fda.gov/scripts/cdrh/cfdocs/cfCFR/CFRSearch.cfm?fr=184.1137 (accessed on 1 March 2021).

19. Available online: https://en.wikipedia.org/wiki/Ammonium_carbonate (accessed on 1 March 2021).
20. Sevastaki, M.; Papadakis, V.M.; Romanitan, C.; Suchea, M.P.; Kenanakis, G. Photocatalytic Properties of Eco-Friendly ZnO Nanostructures on 3D-Printed Polylactic Acid Scaffolds. *Nanomaterials* **2021**, *11*, 168. [CrossRef]
21. Sun, X.; Chen, X.; Deng, Z.; Li, Y. A CTAB-assisted hydrothermal orientation growth of ZnO nanorods. *Mater. Chem. Phys.* **2003**, *78*, 99–104. [CrossRef]
22. Panchakarla, L.S.; Govindaraj, A.; Rao, C.N.R. Formation of ZnO Nanoparticles by the Reaction of Zinc Metal with Aliphatic Alcohols. *J. Clust. Sci.* **2007**, *18*, 660–670. [CrossRef]
23. Zhao, A.; Luo, T.; Chen, L.; Liu, Y.; Li, X.; Tang, Q.; Cai, P.; Qian, Y. Synthesis of ordered ZnO nanorods film on zinc-coated Si substrate and their photoluminescence property. *Mater. Chem. Phys.* **2006**, *99*, 50–53. [CrossRef]
24. Umar, A.; Kim, S.; Im, Y.; Hahn, Y. Structural and optical properties of ZnO micro-spheres and cages by oxidation of metallic Zn powder. *Superlattices Microstruct.* **2006**, *39*, 238–246. [CrossRef]
25. Ma, Q.L.; Xiong, R.; Zhai, B.G.; Huang, Y.M. Water Assisted Conversion of ZnO from Metallic Zinc Particles. *Key Eng. Mater.* **2013**, *538*, 38–41. [CrossRef]
26. Zhang, J.; Sun, L.; Yin, J.; Su, H.; Liao, C.; Yan, C. Control of ZnO morphology via a simple solution route. *Chem. Mater.* **2002**, *14*, 4172–4177. [CrossRef]
27. Suchea, M.; Tudose, I.V.; Vrinceanu, N.; Istrate, B.; Munteanu, C.; Koudoumas, E. Precursor concentration effect on structure and morphology of ZnO for coatings on fabric substrates. *Acta Chim. Iasi* **2013**, *21*, 107–116. [CrossRef]
28. Yin, J.; Gao, F.; Wei, C.; Lu, Q. Water Amount Dependence on Morphologies and Properties of ZnO nanostructures in Double-solvent System. *Sci. Rep.* **2014**, *4*, 3736. [CrossRef]
29. Williamson, G.; Hall, W. X-ray line broadening from filed aluminium and wolfram. *Acta Met.* **1953**, *1*, 22–31. [CrossRef]
30. Rahman, F. Zinc oxide light-emitting diodes: A review. *Opt. Eng.* **2019**, *58*, 010901. [CrossRef]
31. Khranovskyy, V.; Lazorenko, V.; Lashkarev, G.; Yakimova, R. Luminescence anisotropy of ZnO microrods. *J. Lumin.* **2012**, *132*, 2643–2647. [CrossRef]
32. Chen, X.-B.; Huso, J.; Morrison, J.L.; Bergman, L. The properties of ZnO photoluminescence at and above room temperature. *J. Appl. Phys.* **2007**, *102*, 116105. [CrossRef]
33. Serrano, A.; Arana, A.; Galdámez, A.; Dutt, A.; Monroy, B.; Güell, F.; Santana, G. Effect of the seed layer on the growth and orientation of the ZnO nanowires: Consequence on structural and optical properties. *Vacuum* **2017**, *146*, 509–516. [CrossRef]
34. Morkoc, H.; Özgür, Ü. *Zinc Oxide: Fundamentals, Materials and Device Technology*; John Wiley & Sons: Hoboken, NJ, USA, 2008; ISBN 978-3-527-62395-2.
35. Jagadish, C.; Pearton, S. *Zinc Oxide Bulk, Thin Films and Nanostructures*; Elsevier B.V.: New York, NY, USA, 2006; ISBN 978-0-08-044722-3.
36. Leiter, F.H.; Alves, H.R.; Hofstaetter, A.; Hofmann, D.M.; Meyer, B.K. The oxygen vacancy as the origin of a green emission in undoped ZnO. *Phys. Status Solidi B* **2001**, *226*, R4–R5. [CrossRef]
37. Quemener, V.; Vines, L.; Monakhov, E.V.; Svensson, B.G. Evolution of deep electronic states in ZnO during heat treatment in oxygen- and zinc-rich ambients. *Appl. Phys. Lett.* **2012**, *100*, 112108. [CrossRef]
38. Markus Raphael Wagner, Fundamental Properties of Excitons and Phonons in ZnO: A Spectroscopic Study of the Dynamics, Polarity, and Effects of External Fields. 2010. Available online: https://d-nb.info/1064813542/34 (accessed on 1 March 2021).

 *nanomaterials*

Article

Innovative Low-Cost Carbon/ZnO Hybrid Materials with Enhanced Photocatalytic Activity towards Organic Pollutant Dyes' Removal

Petronela Pascariu [1,2,*], Niculae Olaru [1], Aurelian Rotaru [2] and Anton Airinei [1]

[1]   "Petru Poni" Institute of Macromolecular Chemistry, 41A Grigore Ghica Voda Alley, 700487 Iasi, Romania; nolaru@icmpp.ro (N.O.); airineia@icmpp.ro (A.A.)

[2]   Faculty of Electrical Engineering and Computer Science & MANSiD Research Center, Stefan cel Mare University, 13 Str. Universitatii, 720229 Suceava, Romania; aurelian.rotaru@usm.ro

*   Correspondence: dorneanu.petronela@icmpp.ro or pascariu_petronela@yahoo.com

Received: 10 August 2020; Accepted: 15 September 2020; Published: 18 September 2020

Abstract: A new type of material based on carbon/ZnO nanostructures that possesses both adsorption and photocatalytic properties was obtained in three stages: cellulose acetate butyrate (CAB) microfiber mats prepared by the electrospinning method, ZnO nanostructures growth by dipping and hydrothermal methods, and finally thermal calcination at 600 °C in N_2 for 30 min. X-ray diffraction (XRD) confirmed the structural characteristics. It was found that ZnO possesses a hexagonal wurtzite crystalline structure. The ZnO nanocrystals with star-like and nanorod shapes were evidenced by scanning electron microscopy (SEM) measurements. A significant decrease in E_g value was found for carbon/ZnO hybrid materials (2.51 eV) as compared to ZnO nanostructures (3.21 eV). The photocatalytic activity was evaluated by studying the degradation of three dyes, Methylene Blue (MB), Rhodamine B (RhB) and Congo Red (CR) under visible-light irradiation. Therefore, the maximum color removal efficiency (both adsorption and photocatalytic processes) was: 97.97% of MB (C_0 = 10 mg/L), 98.34% of RhB (C_0 = 5 mg/L), and 91.93% of CR (C_0 = 10 mg/L). Moreover, the value of the rate constant (k) was found to be 0.29×10^{-2} min^{-1}. The novelty of this study relies on obtaining new photocatalysts based on carbon/ZnO using cheap and accessible raw materials, and low-cost preparation techniques.

Keywords: carbon/ZnO nanostructures; electrospinning; photocatalyst; photocatalytic activity

---

1. Introduction

A major worldwide problem of modern society is the disposal and treatment of wastewater coming from industrial processes. It is known that about 97% of water is represented by oceans in the form of salty water. This is not appropriate for human consumption or agricultural use, and only less than 3% of water is useful [1]. The quality and quantity of water are the main issues that need to be addressed by finding methods to eliminate contaminants or pollutants195 induce adverse environmental effects, as well as for human health. In addition, the residual liquids containing dyes coming from the textile industry often create severe environmental hazards because of their direct disposal into nearby water bodies. More than 15% of the dyes are lost in wastewater during dyeing operations. This affects the surface esthetic merit of water and reduces light penetration, disturbing aquatic life and hindering photosynthesis [2]. Furthermore, some dyes are either toxic, mutagenic or/and carcinogenic [1].

It is known that ZnO is considered one of the most important oxide semiconductors with a band gap energy of 3–3.37 eV and a large exciton binding energy of 60 meV, having a high capacity to decompose organic pollutants under ultraviolet (UV) irradiation or sunlight exposure [3]. Due to its unique properties, ZnO is widely used for a large variety of applications such as light-emitting diodes, nanolasers, piezo-electric devices, UV-shielding materials, antibacterial agents, field effect transistors, solar cells and gas sensors [4–9]. Moreover, this semiconductor material is considered an excellent photocatalyst for the degradation of some organic dyes in wastewater. Many researchers have been trying to improve the photocatalytic properties of ZnO by doping with various metals (La, Sm, Er, Ce, N, Ag, and so on,) [10–13], combining with other metal oxides (NiO, CeO$_2$, SnO$_2$, CuO, CdO, BaTiO$_3$, NaNbO$_3$, TiO$_2$, Bi$_2$O$_3$, CuFe$_2$O$_4$) [14] or with various carbon-based nanostructures (multi-walled carbon nanotubes (MWCNTs), graphene, graphene oxide) [15–18].

Recently, composite materials based on a combination of metal oxide semiconductor nanomaterials and different types of carbon species have been intensively used as photocatalysts due to their remarkable physico-chemical properties and potential applications in water purification and environmental protection [1,19–21]. In addition, it was demonstrated that the development of materials based on ZnO and carbon leads to an increase in the stability and efficiency of photocatalytic performance [1,19–21]. One of the simplest and cheapest methods of obtaining carbon-based materials is the use of polymer matrices followed by their carbonization at high temperature in N$_2$ atmosphere. The most widely used polymer in obtaining ZnO/carbon-based nanomaterials is polyacrylonitrile (PAN) and is generally used as electrodes for supercapacitors [22–24]. It was shown that carbon nanofibers play an important role in energy conversion and storage, catalysis, sensors, adsorption/separation, and biomedical applications due to its good conductivity and chemical stability, tunable structural flexibility, and low cost [24,25].

The main goal of this study is to point out the remarkable results of carbon/ZnO-based catalysts in photocatalytic degradation, starting from easily accessible and low-cost materials. For this purpose, electrospun fiber mats of cellulose acetate butyrate (CAB) were chosen for growing on them the desired ZnO nanostructures followed by their calcination at 600 °C in N$_2$ atmosphere for 30 min. It is known that CAB is a thermoplastic polymer that softens in the first phases and then follows the degradation process. Moreover, it is a relatively inexpensive and accessible polymer. In this work, we aimed to produce a new type of material based on carbon/ZnO nanostructures that possess both adsorption and photocatalytic properties. To our knowledge, the development of carbon/ZnO nanostructures starting from CAB fiber mats obtained by electrospinning method, followed by ZnO nanostructures growth on them by dipping or hydrothermal method, and finally thermal calcination at 600 °C in N$_2$ atmosphere for 30 min, and then their testing for adsorption/degradation of organic dyes, have not been reported in the literature so far. The details on the structural, morphological, and optical properties of the carbon/ZnO nanostructures were achieved and discussed. Furthermore, the development of new hybrid materials based on carbon/ZnO nanostructures will add valuable insights to scientific research by combining adsorption and photocatalytic processes.

2. Materials and Methods

2.1. Materials

Cellulose acetate butyrate was obtained by commercial sources, Eastman product—CAB 551-0.2—with 52 wt% butyryl content and 2 wt% acetyl content (Mn = 30,000), zinc acetate [Zn(CH$_3$COO)$_2$·2H$_2$O], purchased from Sigma-Aldrich (Taufkirchen, Germany), and ammonia (NH$_3$), purchased from Chemical Company SA, Iasi, Romania. All the chemicals were of reagent grade and were used without further purification.

2.2. Carbon/ZnO Hybrid Nanostructures Preparation

CAB fibers were prepared using the electrospinning method described in detail in previous work [26]. Briefly, the prepared viscous solution of 32% CAB in 2-methoxyethanol as the solvent was transferred in a syringe of the electrospinning setup. The main parameters of the electrospinning process were: high voltage source (25 kV), a 15 cm distance between the needle tip and the collector, and the flow-rate of 0.75 mL/h. Two methods were used in the growth of ZnO nanocrystals in CAB membranes: dipping and hydrothermal, followed by heat treatment at 600 °C for 30 min in N_2 atmosphere to obtain carbon/ZnO hybrid nanostructures. The **M1** sample was obtained using the dipping procedure which consists in successive dippings of the membrane in an ammonium zincate bath with 0.1 M concentration and pH = 11, at room temperature, and then in a hot water bath, at about 97 °C, in 50 repeating cycles. After that, the sample **M1** was thermally treated at 240 °C in the air for 1 h.

The **M2** sample was prepared by the hydrothermal method consisting of: (i) ZnO seeded onto CAB nanofiber mat by 10 dippings; (ii) growth of ZnO nanocrystals by a hydrothermal method in ammonium zincate bath at (96–98 °C) for 3 h, followed by heat treatment at 240 °C in the air for 1 h. The carbon/ZnO hybrid nanostructures **M1 (T)** and **M2 (T)** were developed after calcining of membrane **M1** and **M2** at 600 °C in N_2 atmosphere for 30 min. A representative diagram in preparing the carbon/ZnO hybrid materials is given in Scheme 1.

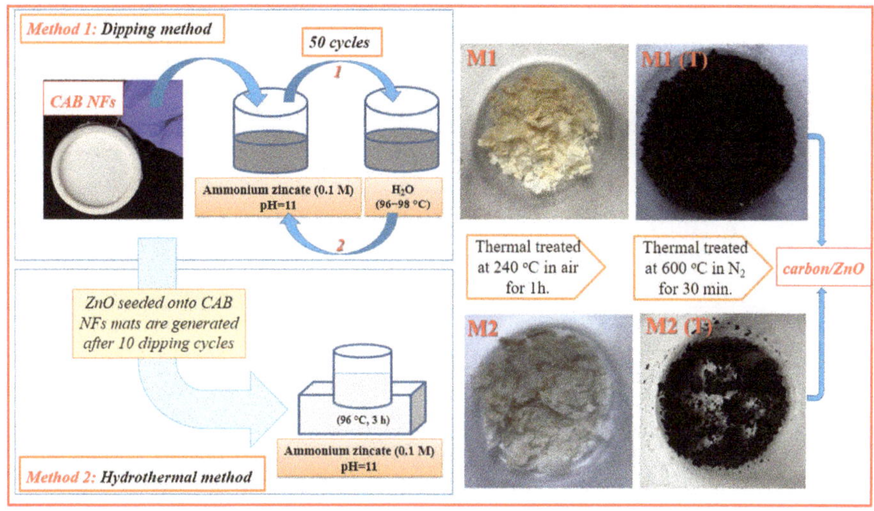

Scheme 1. Preparation of the Carbon/ZnO hybrid nanostructures.

2.3. Characterization of Materials

X-ray diffraction (XRD) analysis of carbon/ZnO hybrid nanostructures as made on a Shimadzu Lab X XRD-6000 diffractometer (Columbia, United States) with CuK_α radiation (λ = 0.15418 nm). The morphological properties of the obtained materials were demonstrated by scanning electron microscopy (SEM), using JEOL JSM 6362LV electron microscope (Japan). A Bruker Fourier transform infrared (FTIR) spectrometer (VERTEX 70, Ettlingen, Germany) equipped with a Deuterated Lanthanum α Alanine doped TriGlycine Sulphate (DLaTGS) detector was used for the analysis of the FTIR spectra of materials. Diffuse reflectance of carbon/ZnO hybrid materials was performed by ultraviolet–visible (UV–Vis) reflectance spectra measured on an Analytik Jena UV-Vis 210 spectrometer (Jena, Germany). Then, the band gap values were obtained using Kubelka–Munk function (KM) and by plotting $[F(R_\infty)h\nu]^2$ vs. $h\nu$.

2.4. Photocatalytic Tests

The adsorption and photocatalytic efficiency of carbon/ZnO hybrid nanostructures have been evaluated by degradation of Methylene Blue (MB), Congo Red (CR), and Rhodamine B (RhB) dye in aqueous solutions under visible light irradiation (100 W tungsten lamp source). More details on the degradation procedure and working conditions have been reported previously [27]. Initially, 5 mg of each material were dispersed in 10 mL of dye solution with an initial concentration of 10 mg/L MB, CR, and 5 mg/L of RhB, respectively. Then, the solutions were stirred in the dark for 2 h to establish an adsorption-desorption equilibrium. The photocatalytic activity of the carbon/ZnO hybrid nanostructures was investigated by photodegradation of MB, CR, and RhB dyes using the same experimental setup and degradation procedure as reported by the authors in a previous work [28]. The UV–Vis absorption profiles for the initial dye solution and after exposure to visible light at various time intervals were obtained using UV-Vis spectrophotometer (SPECORD 210Plus, Analytik Jena, (Jena, Germany). Adsorption capacity (Q_e, mg/g) and removal efficiency (%) for adsorption and degradation of MB were calculated using the following equations [29]:

$$q_e = \frac{(C_0 - C_e) \times V}{m} \times 100, \tag{1}$$

$$\text{Color removal efficiency } (\%) = \frac{C_0 - C_e}{C_0} \times 100, \tag{2}$$

where C_0 is the initial MB concentration (mg/L) and C_e is the MB concentration at the time t (mg/L), m is the catalyst mass (g), and V is the solution volume (L).

3. Results

3.1. X-ray Diffraction (XRD) Characterization

X-ray diffraction (XRD) patterns of **M1 (T)** and **M2 (T)** materials are shown in Figure 1 and confirm the crystalline phase of ZnO with the hexagonal wurtzite structure.

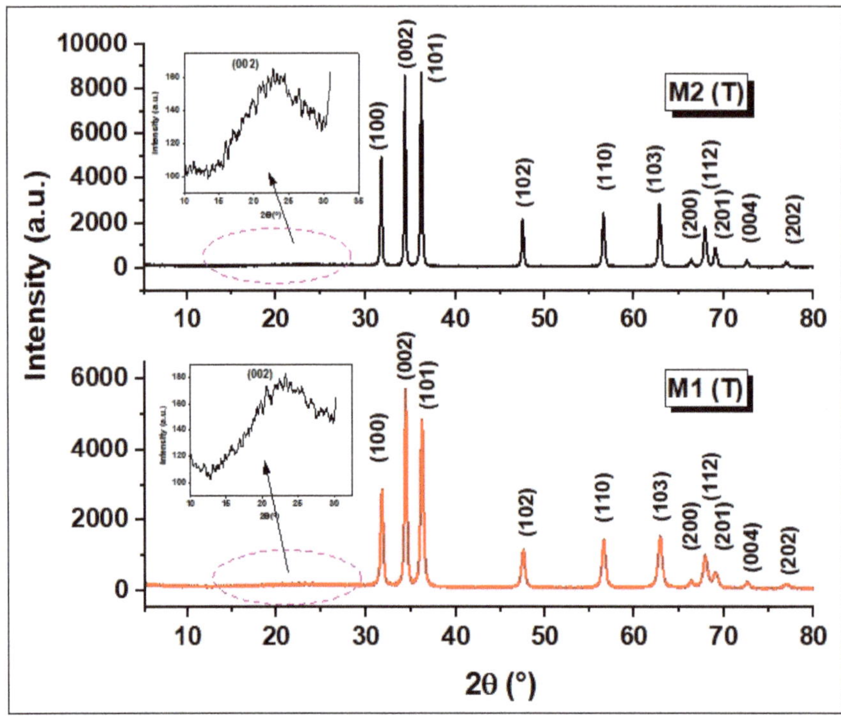

Figure 1. X-ray diffraction (XRD) patterns of **M1 (T)** and **M2 (T)** samples.

The peaks corresponding to this structure are found at 2θ of 31.78° (100), 34.46° (002), 36.26° (101), 47.64° (102), 56.62° (110), 62.90° (103), and 67.10° (112) and belong to pure ZnO structure Joint Committee on Powder Diffraction Standards (JCPDS No. 89-1397). The main parameters that can be deduced from the analysis of X-ray diffractograms are summarized in Table 1, and for their calculation, the diffraction peaks corresponding to the Miller indices (100), (002) and (101) were used. In addition, to estimate the crystallite size (D), the spacing distance between crystallographic planes (d_{hkl}), the lattice parameters a and c, the Zn–O bond length (L) and the microstrain (ε), the authors utilized the equations described in detail in previous work [11].

Table 1. Structural parameters of carbon/ZnO nanostructured materials.

Sample	2θ (°)	d_{hkl} (Å)	D (nm)	Lattice Parameters			ε (%)	L (nm)
				a (Å)	c (Å)	c/a		
M1 (T)	31.78	2.813	22.88	3.249	5.201	1.601	0.564	1.954
	34.46	2.6	20.73					
	36.26	2.475	21.39					
M2 (T)	31.76	2.815	36.13	3.251	5.207	1.602	0.353	1.957
	34.42	2.603	33.17					
	36.24	2.476	30.94					

From the analysis of the lattice parameters a and c presented in Table 1, it can be seen that they do not show significant changes after the carbonization of the organic material, which confirms that the hexagonal wurtzite structure of ZnO is maintained. Besides, the ratio c/a is practically constant, which indicates that the hexagonal wurtzite structure of ZnO structure does not change. Significant changes can be observed for the crystallite size (D) and the microstrain (ε) parameter. The crystallite size values vary between 22.88 nm (corresponding to **M1 (T)** sample) and 36.13 nm (registered for **M2 (T)**), respectively. In addition, a discreet broadening of **M1 (T)** signals was observed which may be ascribed to the presence of a star-like shape of the ZnO crystallites, having a more multidirectional distribution. Moreover, it is well known that a smaller size of crystallites will induce a broadening of the signal. The microstrain (ε) parameter increases from 0.353 corresponding to sample **M2 (T)** to 0.564 for sample **M1 (T)**, probably due to the shape change of the nanostructures and the carbon content of the samples. XRD analysis (Figure 1 (inset)) suggests the presence of carbon in both samples with broad diffraction peaks between 20° and 30°, which was assigned to the (002) lattice planes in the graphitic structure [20]. A significant difference can be observed in the value obtained for crystallites size in XRD compared to those observed in SEM. It is known that the formation of these nanostructures (star-like and nanorod shapes in our case) takes place in two stages: nucleation and growth. In the first stage, small nuclei are formed which, as the reaction progresses, these nuclei grow further to produce star-like and nanorod ZnO crystallites, which are the building blocks for the crystals observed in SEM images [30–34].

3.2. Fourier Transform Infrared (FTIR) Analysis

Figure 2 shows the FTIR spectra of CAB nanofibres and carbon/ZnO corresponding to **M1 (T)** and **M2 (T)** nanostructured materials registered between 370 and 4000 cm^{-1}.

It is known that ZnO has an intense broad band between 420 cm^{-1} and 510 cm^{-1} due to two transverse optical stretching modes of ZnO [35,36]. In our case, two characteristic absorption bands located at 397 cm^{-1} and 497 cm^{-1} were observed corresponding to **M2 (T)** material, as well as an absorption band at 424 cm^{-1} of **M1 (T)**, respectively. The occurrence of these two bands in the FTIR spectrum for sample **M2 (T)** it is due to the nanorod shape nanostructures. Wu et al., [37] state that the transition from 0D nanostructures (nanoparticles) to 1D (nanorod) leads to the appearance of two main absorption maxima in FTIR spectra in this range. The presence of vibration bands at the wavenumbers of 1614 cm^{-1} and 1529 cm^{-1} assigned to the asymmetric stretching vibration and symmetric stretching vibration of C=C bonds indicates the removal of functional groups and the successful carbonization of the new material, as it was confirmed by other authors for similar systems [38]. This aspect is very important since amorphous carbon is known as a very good adsorbent [39]. The absorption band located at 3425 cm^{-1} belongs to the stretching vibration of O–H groups due to the absorbed water on the surface of the carbon/ZnO materials. The bands around 1083 cm^{-1} are associated with bending vibrations of various ether bridges coming from the residual polymeric material.

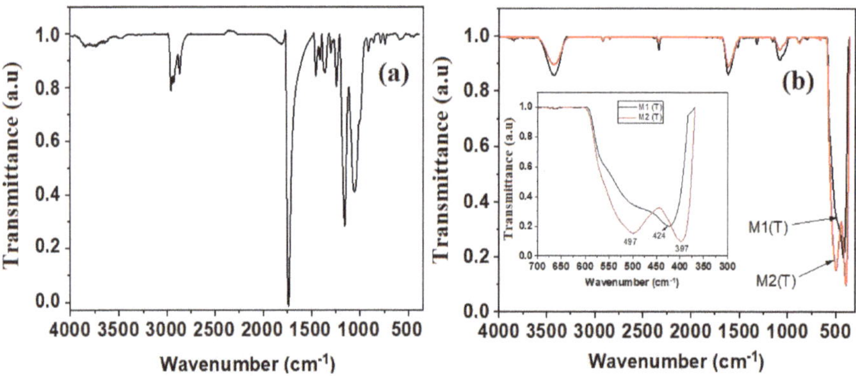

Figure 2. Fourier transform infrared (FTIR) spectra of CAB nanofibres (**a**) and carbon/ZnO; (**b**) for the two types of nanostructures **M1 (T)** and **M2 (T)**, respectively.

3.3. Morphological Characterization

The SEM image of the CAB microfiber obtained immediately after the electrospinning process is shown in Figure 3.

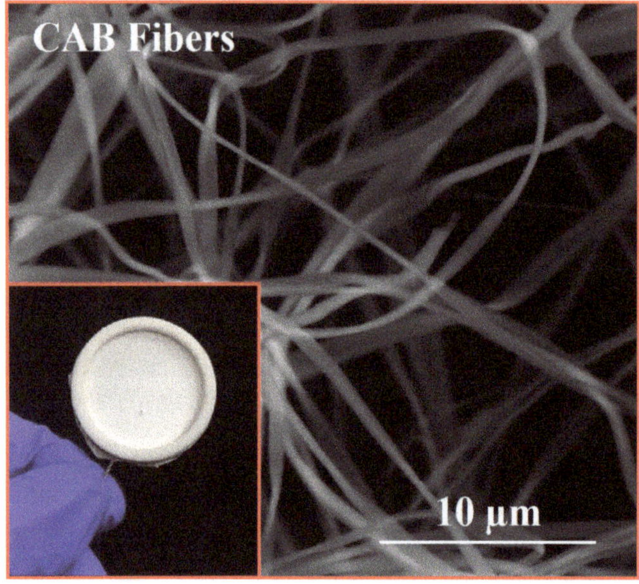

Figure 3. Scanning electron microscope (SEM) image of cellulose acetate butyrate (CAB) microfiber mats obtained by the electrospinning method from a mixture of 32% polymer and 2-methoxyethanol.

This micrograph confirms the formation of a membrane with uniform and smooth microfibers having dimensions of 1 μm. After this, star-shaped crystals and nanorods were grown on these membranes by dipping and hydrothermal methods, followed by calcination at 240 °C in the air for 1 h. SEM images shown in Figure 4 for hybrid CAB/ZnO nanostructures obtained by the dipping method indicate a structure composed of ZnO nanocrystals with a star-like shape. It can be observed that the same structure was maintained after calcination at 600 °C in N_2 atmosphere for 30 min for **M1 (T)** nanostructure (Figure 4).

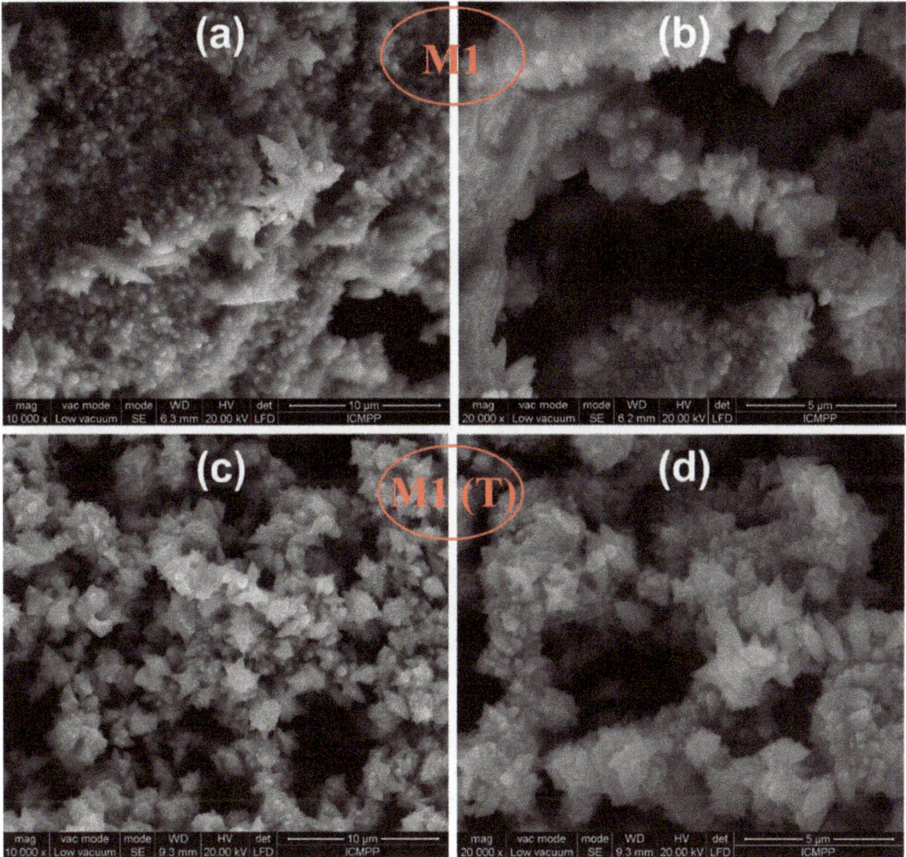

Figure 4. SEM images of CAB/ZnO hybrid nanostructures obtained by the dipping method after thermal treating at 240 °C in the air for 1 h ((**a**) and (**b**) for **M1**), and carbon/ZnO after thermally treating at 600 °C in N_2 for 30 min ((**c**) and (**d**) for **M1 (T)**).

The materials obtained by the hydrothermal method (**M2** and **M2 (T)**) show a nanorod type structure with an average diameter of about 700 nm and lengths around 5 μm according to the SEM images represented in Figure 5.

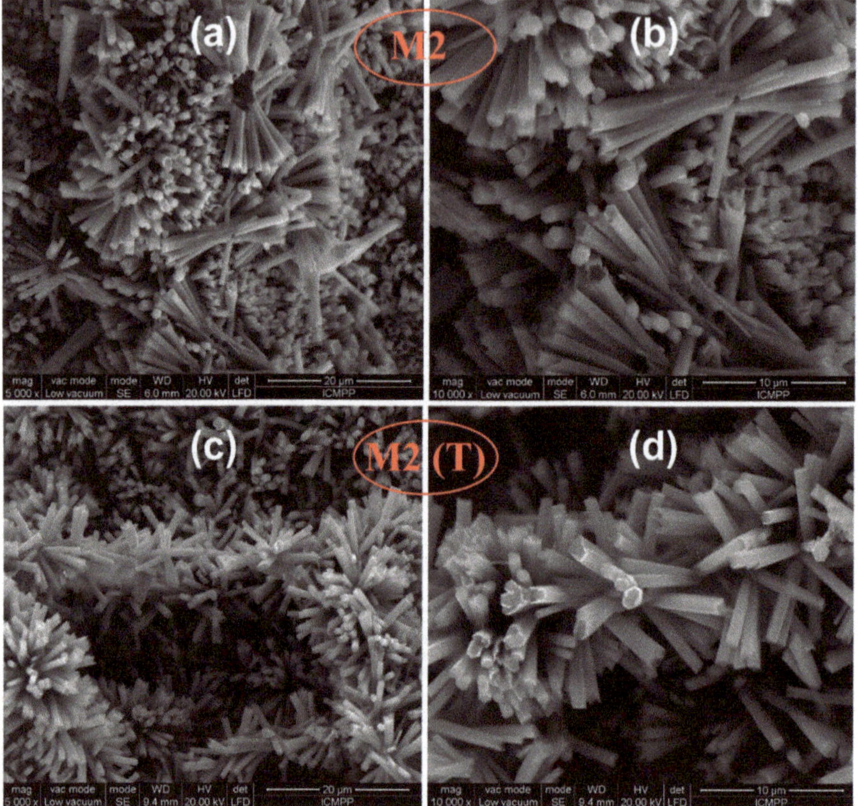

Figure 5. SEM analysis of CAB/ZnO hybrid nanostructures obtained by the hydrothermal method after thermal treating at 240 °C in the air for 1 h ((**a**) and (**b**) for **M2**), and carbon/ZnO after thermally treating at 600 °C in N_2 for 30 min ((**c**) and (**d**) for **M2 (T)**).

3.4. Optical Properties

The most important parameter that significantly influences the photodegradation process is represented by the energy band gap of materials. The value of this parameter was assessed by UV–Vis reflectance experiments, followed by applying the Kubelka–Munk equation (Equation (3)) and Tauc relation (Equation (4)) [40].

$$F(R_\infty) = \frac{(1 - R_\infty)^2}{2R_\infty},\tag{3}$$

where $F(R_\infty)$ is the so-called remission or Kubelka–Munk function and R_∞ is the reflectance of the samples.

$$[F(R_\infty)h\nu]^2 = A\big(h\nu - E_g\big),\tag{4}$$

where A is a constant, E_g is the optical band gap of the material.

The energy band gap values of carbon/ZnO nanostructures were obtained by plotting $[F(R_\infty)h\nu]^2$ versus $h\nu$ and extrapolating the linear portion of the absorption edge to find the intercept with photon energy axis as shown in Figure 6.

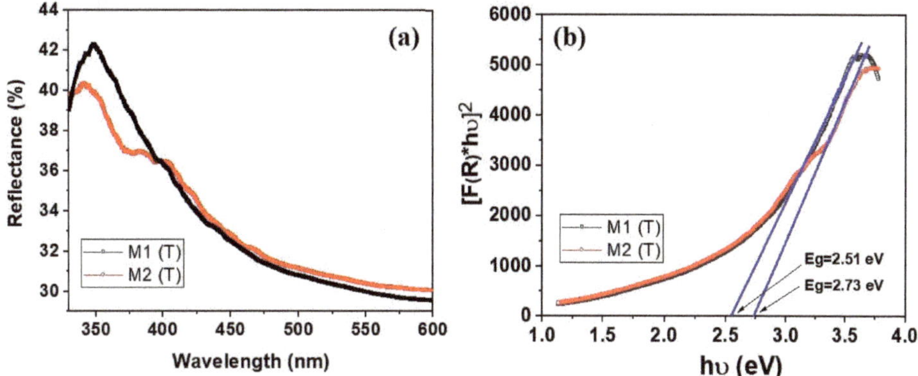

Figure 6. Optical properties of carbon/ZnO nanostructures (**a**) diffuse reflectance ultraviolet–visible (UV–Vis) spectra and (**b**) plot of $[F(R)h\nu]^2$ versus hν and band gap determination.

A significant decrease in E_g value was observed for carbon/ZnO hybrid materials. Thus, for **M1 (T)** the E_g was found to be 2.51 eV, while for **M2 (T)** 2.73 eV, respectively. These values are lower compared to those obtained for CAB/ZnO (3.21 eV and 3.31 eV) nanostructures reported in our previous works [26]. This decrease of E_g could be ascribed to the enhanced conductivity, confirmed by other authors for similar systems [41]. It can be seen that the presence of carbon in ZnO nanostructures leads to a change in the electronic energy levels. For example, similar results were obtained for hybrid RGO-ZnO where E_g decreases to 2.16 eV as compared to pure ZnO (3.06 eV) [42]. Another study reported by Rahimi et al., [15] showed that the E_g value decreases from 3.2 eV (ZnO) to 2.8 eV for ZnO nanorod/graphene quantum dot composites, respectively. The authors associate this phenomenon to the formation of Zn–O–C or Zn–C chemical bonds in the composites obtained.

3.5. Photoluminescence Study

The analysis of the photoluminescence properties is closely related to the photocatalytic properties of the developed catalysts and help us to understand the recombination processes of the photogenerated electron-hole pairs. Therefore, the emission spectra obtained under 300 nm and 320 nm excitation wavelengths are presented in Figure 7.

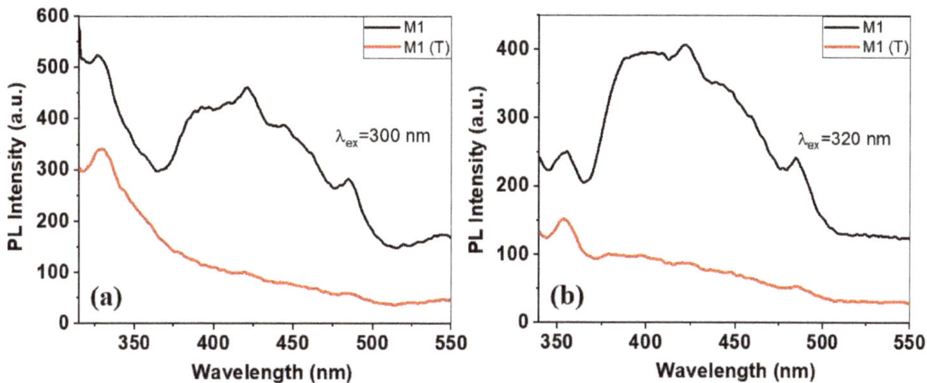

Figure 7. Photoluminescence spectra of CAB/ZnO and Carbon/ZnO nanostructures as a function of excitation wavelength 300 (**a**), and 320 nm (**b**).

It can be seen that the emission spectra corresponding to sample **M1** (CAB/ZnO) show several emission bands at 327 nm, 391 nm, 421 nm, 444 nm, and 484 nm, respectively. The UV emission bands from 327 nm (Figure 7a) and 350 nm (Figure 7b) can be assigned the near band edge (NBE) emission, and may be due to free exciton recombination [43]. It is known that the emission bands in the visible spectrum are due to different intrinsic defects of ZnO nanostructures, which include oxygen vacancies (V_O), zinc vacancies (V_{Zn}), oxygen interstitials (O_i), zinc interstitials (Zn_i) and oxygen antisites (O_{Zn}) [27].

The emission spectrum of carbon/ZnO sample provides weak photoluminescence compared to sample **M1**. This means that the absorbed light is used efficiently in generating hole-electron pairs, without losing in the form of photoluminescence. The band located in the blue region practically disappeared, and the bands at 327 (Figure 7a), 350 nm (Figure 7b) and 484 nm become very weak. According to other studies [15], this large decrease of photoluminescence of carbon/ZnO nanostructures may indicate a large decrease in the radiative recombination rate of electron-hole pairs.

3.6. Adsorption/Photocatalytic Properties

3.6.1. Adsorption/Photocatalytic Properties of Carbon/ZnO Hybrid Nanostructures for Degradation of Organic Pollutants

In the first stage of this study, the degradation efficiency of rhodamine B ($C_0 = 5$ mg/L) for the starting samples (**M1** and **M2**) and the calcined samples (**M1 (T)** and **M2 (T)**) in N_2 was performed. The blank test (without catalyst) was initially evaluated after 4 h and showed that the intensity of the absorption band of RhB decreases slightly, yielding 1.39% in dye degradation. Figure 8 shows the evolution of the absorption spectra of all materials after adsorption for two hours to establish the adsorption/desorption equilibrium of dye on the photocatalyst surface, followed by the degradation between 4 and 20 h depending on the efficiency of the samples.

From the analysis of the samples, it was noticed that for the samples **M1** and **M2** the adsorption process was very small (5.56% for **M1** and 1.08% corresponding to **M2**). A significant increase in the adsorption process occurs after the carbonization of materials, yielding adsorption efficiency between 89.61% (**M1 (T)**) and 46.59% (**M2 (T)**), respectively. This increase was attributed to the inclusion of carbon in the newly developed hybrid materials. It is known that carbon-based materials lead to an increase in adsorption, conductivity, as well as a decrease in the energy band gap [25]. The most outstanding result, which cumulates both the adsorption/photocatalytic processes, was registered for **M1 (T)** with an efficiency of up to 98.34%.

In the next part of this study, the effect of the initial MB dye concentration on the **M1 (T)** nanostructure activity was investigated. To assess each contribution, adsorption and photodegradation, measurements for five initial dye concentrations (7, 10, 13, 17, and 20 mg/L) were performed. Figure 9 shows that the color removal efficiency in the adsorption process increased with the decrease in the initial dye concentration. Initially, it can suggest that this process is apparently significant, but after calculating, the adsorption constant Q_e (mg/g) for all concentrations was the same (12–13 mg/g) for all samples. Under these conditions, in the next part of this work, the photodegradation of MB dye was evaluated without taking into account the adsorption process.

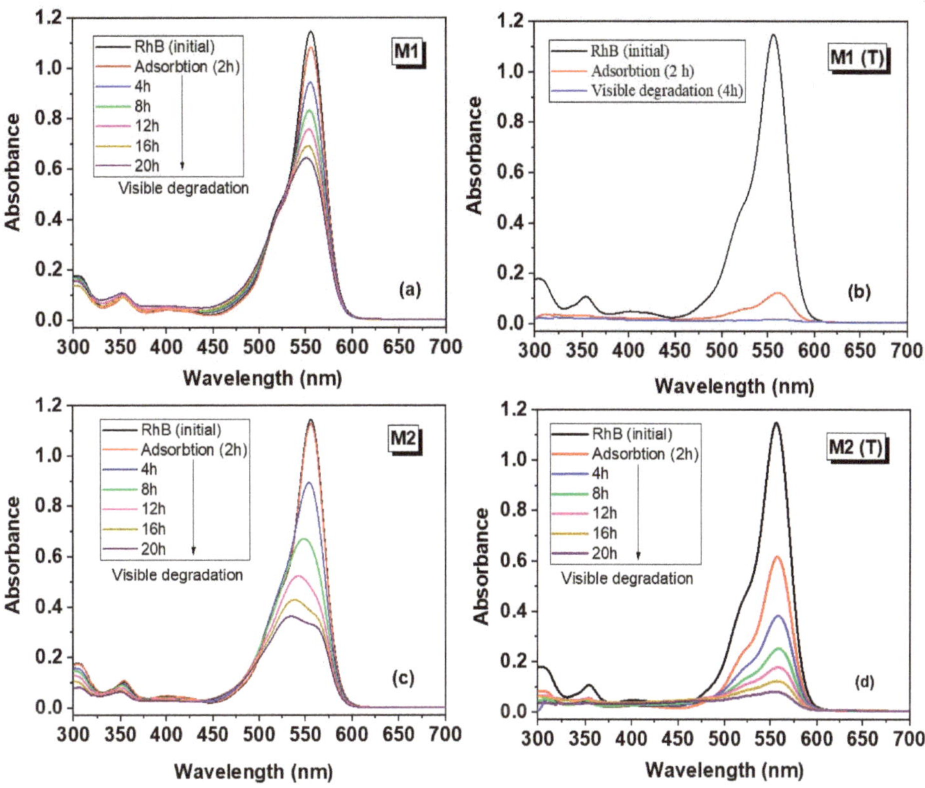

Figure 8. UV–Vis absorption spectra of Rhodamine B (RhB) recorded after adsorption (2h) (in dark) and photodegradation process under visible light irradiation (up to 20 h)) for the materials obtained by dipping method (**M1**, **M1 (T)**) and hydrothermal method (**M2**, **M2 (T)**): (**a,c**) before calcination and (**b,d**) after calcination.

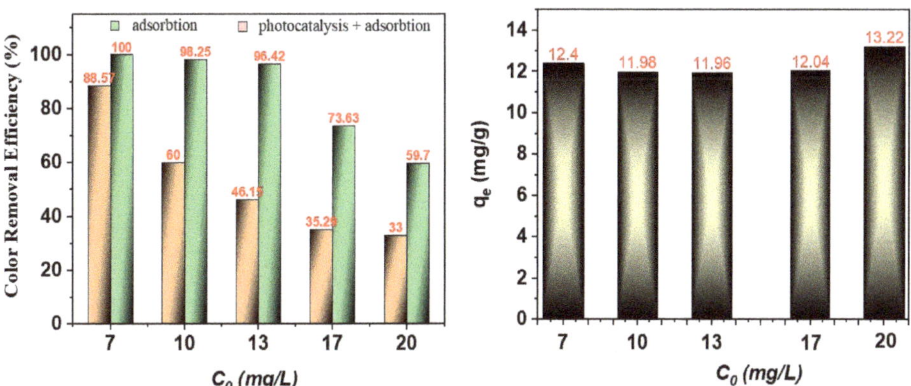

Figure 9. Color removal efficiency after adsorption and photodegradation (**left**), and the maximum adsorption capacity of **M1 (T)** catalyst (**right**) for Methylene Blue (MB) dye.

3.6.2. Photocatalytic Activity of Carbon/ZnO Hybrid Nanostructures for Degradation of Methylene Blue (MB) Dye

Figure 10a,b show the evolution of the UV–Vis absorption spectra for MB dye degradation in presence of both catalysts (**M1 (T)** and **M2 (T)**) under visible light irradiation for 4 h (without previous the 2 h adsorption process).

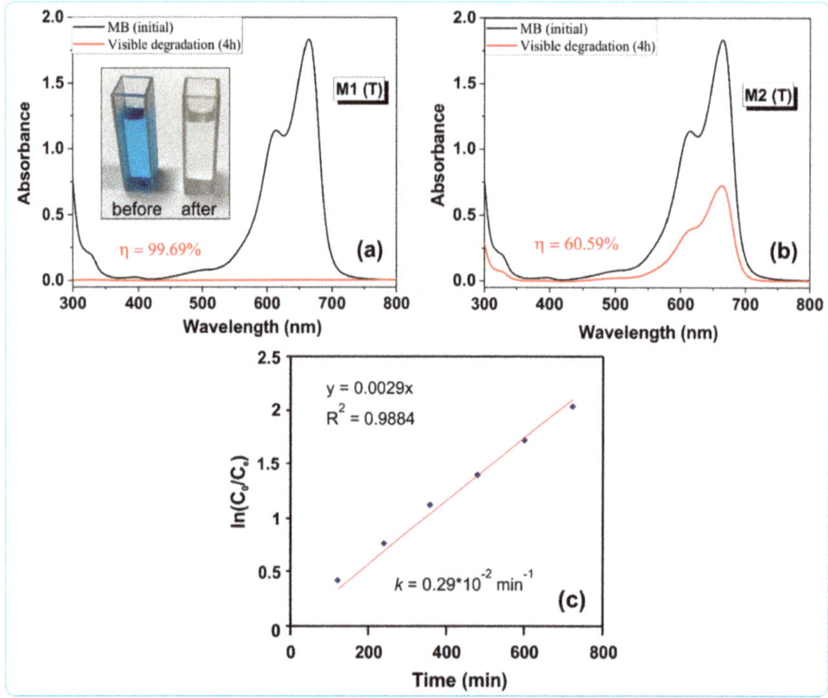

Figure 10. UV–Vis absorption spectra after visible light irradiation for 4 h in the presence of sample **M1 (T)** (a) and **M2 (T)** (b), and the degradation kinetics for **M1 (T)** sample (c).

It was observed that after 4 h of visible light irradiation the absorption band at 665 nm decreases to almost 0, reaching a maximum efficiency of 99.69% for sample **M1 (T)**. The **M2 (T)** sample reveals a slower decrease in the degradation efficiency, yielding 60.59%. We consider that this difference between the values of the photocatalytic degradation efficiency would be due to the different shapes and the presence of carbon in the nanostructures, giving a lower value of the band gap for the **M1 (T)** sample.

Quantitative estimation of degradation kinetics of MB dye was performed using a pseudo-first-order kinetics model according to the following equation: $\ln(C_0/C_t) = kt$, C_0 is concentration of dye solution before irradiation, C_t is concentration of dye solution after t minutes of irradiation, and k is the pseudo-first-order rate constant. The value of the reaction constant for sample **M1 (T)** was calculated by plotting $\ln(C_0/C_t)$ versus irradiation time t (see Figure 10c) and was found to be 0.29×10^{-2} min^{-1} with the value $R^2 = 0.9884$ attributed to a pseudo first-order reaction kinetics.

To demonstrate the adsorption/photocatalytic properties of the new carbon/ZnO hybrid nanostructures it was performed experiments in photocatalytic degradation of MB (initial concentration 10 mg/L), RhB (5 mg/L) and CR (10 mg/L) as a test reaction. Very good results were recorded for the degradation of all dyes tested with the following maximum color removal efficiency (both adsorption and adsorption + photocatalytic processes after 4 h of irradiation): 97.97% for MB (C_0 = 10 mg/L), 98.34% for RhB (C_0 = 5 mg/L), and 91.93% for CR (C_0 = 10 mg/L), respectively (Figure 11).

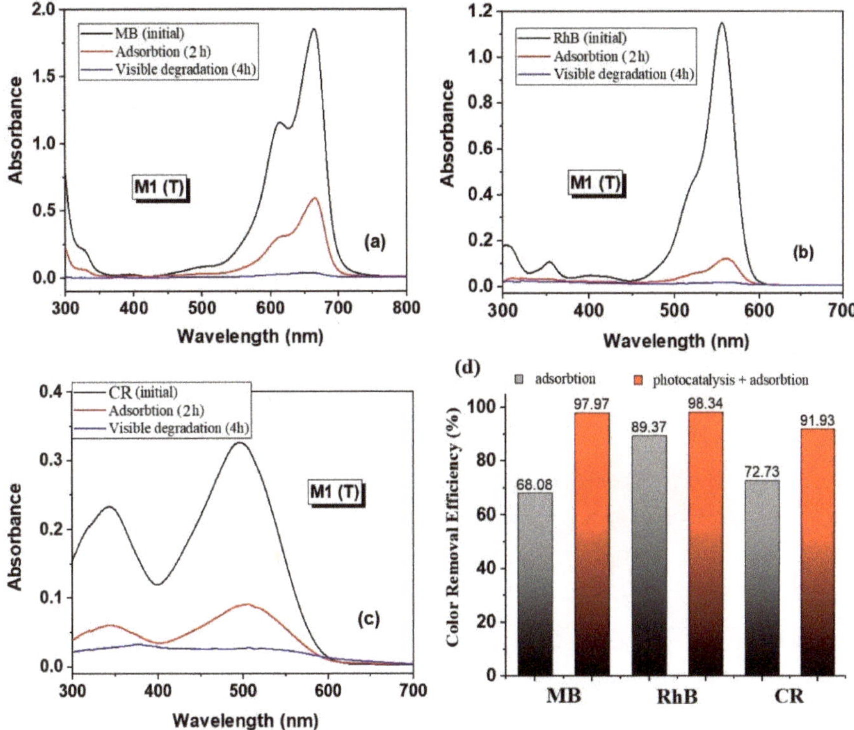

Figure 11. UV–Vis absorption spectra after adsorption (2h) and visible light irradiation (4h) in the presence of sample **M1 (T)** for degradation of MB (C_0 = 10 mg/L) (**a**); RhB (C_0 = 5 mg/L) (**b**); Congo Red (CR) (C_0 = 10 mg/L) (**c**), and color removal efficiency of all dyes degradation for sample **M1 (T)** (**d**).

Therefore, it can be stated that these materials could be employed as promising low-cost photocatalysts with impressive efficiency for potential applications in water purification and environmental protection. Under the given conditions—visible light irradiation at low power (a 100 W tungsten with a power of 102.74 kJ·m^{-2}·h^{-1}), a moderate amount of catalyst (0.5 g/L) and 4 h degradation process—the newly obtained materials present an outstanding response towards organic dyes degradation, with a removal efficiency of 91.93%, 97.97% and 98.34%, depending on the type of dye.

Table 2 reveals the photocatalytic activities represented by the values of the reaction rate constant k (min^{-1}) or degradation efficiency (%) for the degradation of different dyes in the presence of ZnO/carbon-based catalysts. As can be seen, good results were found for these materials based on different carbon nanostructures (reduced graphene oxide, graphene quantum dot, graphene oxide, carbon nanofibers, carbon) [15–18,44]. All authors reported an improvement in photocatalytic activity for these composite materials as compared to ZnO. Instead, the materials analyzed in this study showed enhanced photocatalytic efficiency after 4 h under visible light irradiation at low intensity visible light in the degradation of all dyes (MB, RhB, and CR).

Table 2. Photocatalytic activities of carbon/ZnO nanostructured materials.

Photocatalyst Type	Type and Concentration of Dye	Amount of Photocatalyst (g/L)	Light Source	Reaction Rate Constant k (min^{-1})	η (%)	Ref.
ZnO/RGO	MB (10 mg/L)	1.25	UV (100 W)	0.0395	–	[17]
ZnO/GQDs	MB (2.5×10^{-5} M/L)	–	250 W	0.018 (150 min)	–	[15]
ZnO/GO	Basic Fuchsin (20 mg/L)	0.2	UV	0.00845 (300 min)	92.5	[44]
ZnO/Graphene	MO (5×10^{-5} M/L)	0.08	VIS (300 W)	0.0116	–	[16]
ZnO/CNFs	RhB (10 mg/L)	1	UV (50 W)	0.07533	–	[18]
Carbon/ZnO	MB (10 mg/L)	0.5	Vis (100 W)	0.002 (240 min)	97.97	This work
	RhB (5 mg/L)			–	98.34	
	CR (10 mg/L)			–	91.93	

According to the above results, a mechanism has been proposed to explain the improvement of the photocatalytic efficiency of the carbon/ZnO nanostructures as compared to pure ZnO (Figure 12).

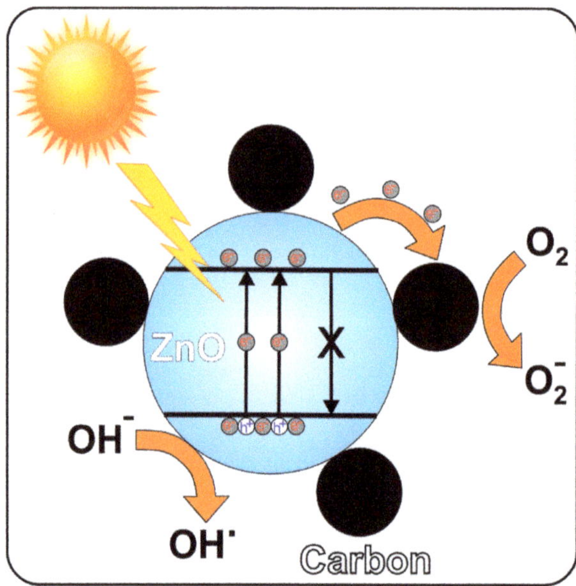

Figure 12. Proposed mechanisms of the photocatalysis of the carbon/ZnO nanostructures.

The degradation mechanism takes into account the cooperative or synergetic effects between the carbon generated during calcination and zinc oxide (Figure 12).

During photon excites, electron hole pairs are generated in the ZnO valence band. These excited electrons will move in the conduction band of ZnO and then diffuse toward the surface of the carbon particles. The holes excess in the valence band will migrate to the surface on ZnO, where they react with water molecules or hydroxyl ions to generate active species of hydroxyl radicals (OH). This method suggests that the photogenerated electrons and holes were effectively separated. Moreover, the good separation of the photogenerated electrons and holes in the carbon/ZnO nanostructures is supported by the photoluminescence investigations of ZnO and carbon/ZnO. According to Figure 7, carbon/ZnO nanostructures revealed weaker emission intensity compared to ZnO. This aspect suggests that the recombination of the photogenerated charge carrier was highly inhibited in the carbon/ZnO nanostructures. The efficient charge separation could induce the increase of the charge carriers' lifetime by enhancing the efficiency of the interfacial charge transfer of the adsorbed substrates. This discussion is also supported by other studies regarding similar systems [17,18].

4. Conclusions

Carbon/ZnO nanostructures were obtained in three stages: CAB microfiber mats were prepared by the electrospinning method, ZnO nanostructures were grown by dipping and hydrothermal methods, followed by thermal calcination at 600 °C in N_2 atmosphere for 30 min. XRD measurements of photocatalysts confirmed a hexagonal wurtzite crystalline structure of ZnO, as well as the presence of carbon with (002) lattice planes. SEM measurements showed the formation of nanostructures with star-like and nanorod shapes. The E_g value decreased significantly for carbon/ZnO hybrid materials (2.51 eV) as compared to ZnO nanostructures (3.21 eV). The photocatalytic efficiency for degradation of Methylene Blue (MB), Rhodamine B (RhB) and Congo Red (CR) dyes under visible-light irradiation has been improved as compared to ZnO. These new materials showed an improvement of the photocatalytic degradation efficiency for the RhB dye with approximately 80% as compared to the ZnO (control samples). The carbon/ZnO hybrid materials recorded a color removal efficiency (adsorption/photocatalytic process) between 91% and 98%, depending on the type of dye. All the experiments were performed under friendly environmental conditions: visible light irradiation at low power and a moderate amount of catalyst (0.5 g/L). Moreover, the value of the rate constant was found for this material to be 0.29×10^{-2} min^{-1}. Therefore, the prepared carbon/ZnO materials from easily accessible and low-cost materials together with their impressive performance place them among photocatalysts for practical applications in wastewater purification.

Author Contributions: Conceptualization; Data curation; Formal analysis; Investigation; Validation; Visualization, Validation, Writing-review and editing; P.P.; Conceptualization; Resources; Supervision; N.O.; Data curation, Supervision, A.R.; Resource, Visualization and Supervision A.A. All authors have read and agreed to the published version of the manuscript.

Funding: This work was supported by contract no. 18PFE/16.10.2018 funded by the Ministry of Research and Innovation within Program 1—Development of national research and development system, Subprogram 1.2—Institutional Performance—RDI excellence funding projects

Conflicts of Interest: The authors declare no conflict of interest.

References

1. Pascariu, P.; Homocianu, M. ZnO-based ceramic nanofibers: Preparation, properties and applications. *Ceram. Int.* **2019**, *45*, 11158–11173. [CrossRef]
2. Singh, K.; Kumar, P.; Srivastava, R. An overview of textile dyes and their removal techniques: Indian perspective. *Pollut. Res.* **2017**, *36*, 790–797.
3. Djurisic, A.B.; Chen, X.; Leung, Y.H.; Ng, A.M.C. ZnO nanostructures: Growth, properties and applications. *J. Mater. Chem.* **2012**, *22*, 6526–6535. [CrossRef]
4. Bolink, H.J.; Coronado, E.; Repetto, D.; Sessolo, M. Air stable hybrid organic-inorganic light emitting diodes using ZnO as the cathode. *Appl. Phys. Lett.* **2007**, *91*, 223501. [CrossRef]
5. Liao, Y.J.; Cheng, C.W.; Wu, B.H.; Wang, C.Y.; Chen, C.Y.; Gwo, S.; Chen, L.J. Low threshold room-temperature UV surface plasmon polariton lasers with ZnO nanowires on single-crystal aluminum films with Al_2O_3 interlayers. *RSC Adv.* **2019**, *9*, 13600–13607. [CrossRef]
6. Yoon, C.; Jeon, B.; Yoon, G. Development of Al foil-based sandwich-type ZnO piezoelectric nanogenerators. *AIP Adv.* **2020**, *10*, 045018. [CrossRef]
7. Kumar, R.; Umar, A.; Kumar, G.; Nalwa, H.S. Antimicrobial properties of ZnO nanomaterials: A review. *Ceram. Int.* **2017**, *43*, 3940–3961. [CrossRef]
8. Saha, J.K.; Bukke, R.N.; Mude, N.N.; Jang, J. Remarkable stability improvement of ZnO TFT with Al_2O_3 gate insulator by yttrium passivation with spray pyrolysis. *Nanomaterials* **2020**, *10*, 976. [CrossRef]
9. Gao, G.; Yu, L.; Vinu, A.; Shapter, J.G.; Batmunkh, M.; Shearer, C.J.; Yin, T.; Huang, P.; Cui, D. Synthesis of ultra-long hierarchical ZnO whiskers in a hydrothermal system for dye-sensitised solar cells. *RSC Adv.* **2016**, *6*, 109406–109413. [CrossRef]

10. Pascariu, P.; Cojocaru, C.; Samoila, P.; Airinei, A.; Olaru, N.; Rusu, D.; Rosca, I.; Suchea, M. Photocatalytic and antimicrobial activity of electrospun ZnO:Ag nanostructures. *J. Alloys Compd.* **2020**, *834*, 155144. [CrossRef]

11. Pascariu, P.; Cojocaru, C.; Olaru, N.; Samoila, P.; Airinei, A.; Ignat, M.; Sacarescu, L.; Timpu, D. Novel rare earth (RE-La, Er, Sm) metal doped ZnO photocatalysts for degradation of Congo-Red dye: Synthesis, characterization and kinetic studies. *J. Environ. Manag.* **2019**, *239*, 225–234. [CrossRef] [PubMed]

12. Lang, J.; Wang, J.; Zhang, Q.; Li, X.; Han, Q.; Wei, M.; Sui, Y.; Wang, D.; Yang, J. Chemical precipitation synthesis and significant enhancement in photocatalytic activity of Ce-doped ZnO nanoparticles. *Ceram. Int.* **2016**, *42*, 14175–14181. [CrossRef]

13. Macías-Sánchez, J.J.; Hinojosa-Reyes, L.; Caballero-Quintero, A.; De la Cruz, W.; Ruiz-Ruiz, E.; Hernández-Ramírez, A.; Guzmán-Mar, J.L. Synthesis of nitrogen-doped ZnO by sol–gel method: Characterization and its application on visible photocatalytic degradation of 2,4-D and picloram herbicides. *Photochem. Photobiol. Sci.* **2015**, *14*, 536–542. [CrossRef]

14. Adnan, M.A.M.; Julkapli, N.M.; Hamid, S.B.A. Review on ZnO hybrid photocatalyst: Impact on photocatalytic activities of water pollutant degradation. *Rev. Inorg. Chem.* **2016**, *36*, 77–104.

15. Rahimi, K.; Yazdani, A.; Ahmadirad, M. Facile preparation of zinc oxide nanorods surrounded by graphene quantum dots both synthesized via separate pyrolysis procedures for photocatalyst application. *Mat. Res. Bull.* **2018**, *98*, 148–154. [CrossRef]

16. Xu, J.; Cui, Y.; Han, Y.; Hao, M.; Zhang, X. ZnO–graphene composites with high photocatalytic activities under visible light. *RSC Adv.* **2016**, *6*, 96778–96784. [CrossRef]

17. Xu, S.; Fu, L.; Pham, T.S.H.; Yu, A.; Han, F.; Chen, L. Preparation of ZnO flower/reduced grapheme oxide composite with enhanced photocatalytic performance under sunlight. *Ceram. Int.* **2015**, *41*, 4007–4013. [CrossRef]

18. Mu, J.; Shao, C.; Guo, Z.; Zhang, Z.; Zhang, M.; Zhang, P.; Chen, B.; Liu, Y. High Photocatalytic activity of ZnO-Carbon nanofiber heteroarchitectures. *ACS Appl. Mater. Interfaces* **2011**, *3*, 590–596. [CrossRef]

19. Batmunkh, M.; Biggs, M.J.; Shapter, J.G. Carbonaceous dye-sensitized solar cell photoelectrodes. *Adv. Sci.* **2015**, *2*, 1400025. [CrossRef]

20. Hu, C.; Lu, T.; Chen, F.; Zhang, R. A brief review of graphene–metal oxide composites synthesis and applications in photocatalysis. *J. Chin. Adv. Mater. Soc.* **2013**, *1*, 21–39. [CrossRef]

21. Concina, I.; Ibupoto, Z.H.; Vomiero, A. Semiconducting metal oxide nanostructures for water splitting and photovoltaics. *Adv. Energy Mater.* **2017**, *7*, 1700706. [CrossRef]

22. Shi, S.; Zhuang, X.; Cheng, B.; Wang, X. Solution blowing of ZnO nanoflake-encapsulated carbon nanofibers as electrodes for supercapacitors. *J. Mater. Chem. A* **2013**, *1*, 13779–13788. [CrossRef]

23. Samuel, E.; Joshi, B.; Kim, M.W.; Kim, Y.I.; Swihart, M.T.; Yoon, S.S. Hierarchical zeolitic imidazolate framework-derived manganese-doped zinc oxide decorated carbon nanofiber electrodes for high performance flexible supercapacitors. *Chem. Eng. J.* **2019**, *371*, 657–665. [CrossRef]

24. Zhang, L.; Aboagye, A.; Kelkar, A.; Lai, C.; Fong, H. A review: Carbon nanofibers from electrospun polyacrylonitrile and their applications. *J. Mater. Sci.* **2014**, *49*, 463–480. [CrossRef]

25. Lee, B.S.; Yu, W.R. Electrospun carbon nanofibers as a functional composite platform: A review of highly tunable microstructures and morphologies for versatile applications. *Funct. Compos. Struct.* **2020**, *2*, 012001. [CrossRef]

26. Pascariu, P.; Olaru, L.; Matricala, A.L.; Olaru, N. Photocatalytic activity of ZnO nanostructures grown on electrospun CAB ultrafine fibers. *Appl. Surf. Sci.* **2018**, *455*, 61–69. [CrossRef]

27. Pascariu, P.; Airinei, A.; Olaru, N.; Olaru, L.; Nica, V. Photocatalytic degradation of Rhodamine B dye using ZnO–SnO$_2$ electrospun ceramic nanofibers. *Ceram. Int.* **2016**, *42*, 6775–6781. [CrossRef]

28. Cojocaru, C.; Pascariu Dorneanu, P.; Airinei, A.; Olaru, N.; Samoila, P.; Rotaru, A. Design and evaluation of electrospun polysulfone fibers and polysulfone/NiFe$_2$O$_4$ nanostructured composite as sorbents for oil spill cleanup. *J. Taiwan Inst. Chem. Eng.* **2017**, *70*, 267–281. [CrossRef]

29. Thamer, B.M.; El-Hamshary, H.; Al-Deyab, S.S.; El-Newehy, M.H. Functionalized electrospun carbon nanofibers for removal of cationic dye. *Arab. J. Chem.* **2019**, *12*, 747–759. [CrossRef]

30. Yun, S.I.; Kim, S.H.; Kim, D.W.; Kim, Y.A.; Kim, B.H. Facile preparation and capacitive properties of low-cost carbon nanofibers with ZnO derived from lignin and pitch as supercapacitor electrodes. *Carbon* **2019**, *149*, 637–645. [CrossRef]

31. Samanta, P.K.; Bandyopadhyay, A.K. Chemical growth of hexagonal zinc oxide nanorods and their optical properties. *Appl. Nanosci.* **2012**, *2*, 111–117. [CrossRef]
32. Ge, M.Y.; Wu, H.P.; Niu, L.; Liu, J.F.; Chen, S.Y.; Shen, P.Y.; Zeng, Y.W.; Wang, Y.W.; Zhang, G.Q.; Jiang, J.Z. Nanostructured ZnO: From monodisperse nanoparticles to nanorods. *J. Cryst. Growth* **2007**, *305*, 162–166. [CrossRef]
33. Seow, Z.L.S.; Wong, A.S.W.; Thavasi, V.; Jose, R.; Ramakrishna, S.; Ho, G.W. Controlled synthesis and application of ZnO nanoparticles, nanorods and nanospheres in dye-sensitized solar cells. *Nanotechnology* **2009**, *20*, 045604. [CrossRef] [PubMed]
34. Lee, T.H.; Sue, H.J.; Cheng, X. Solid-state dye-sensitized solar cells based on ZnO nanoparticle and nanorod array hybrid photoanodes. *Nanoscale Res. Lett.* **2011**, *6*, 517. [CrossRef]
35. Anzlovar, A.; Orel, Z.C.; Kogej, K.; Zigon, M. Polyol-mediated synthesis of zinc oxide nanorods and nanocomposites with poly(methyl methacrylate). *J. Nanomater.* **2012**, *2012*, 760872. [CrossRef]
36. Hayashi, S.; Nakamori, N.; Kanamori, H. Generalized theory of average dielectric constant and its application to infrared absorption by ZnO small particles. *J. Phys. Soc. Jpn.* **1979**, *46*, 176–183. [CrossRef]
37. Wu, L.; Wu, Y.; Pan, X.; Kong, F. Synthesis of ZnO nanorod and the annealing effect on its photoluminescence property. *Opt. Mater.* **2006**, *28*, 418–422. [CrossRef]
38. Zhao, Q.; Xie, H.; Ning, H.; Liu, J.; Zhang, H.; Wang, L.; Wang, X.; Zhu, Y.; Li, S.; Wu, M. Intercalating petroleum asphalt into electrospun ZnO/Carbon nanofibers as enhanced free-standing anode for lithium-ion batteries. *J. Alloys Compd.* **2018**, *737*, 330–336. [CrossRef]
39. García-Díaz, I.; López, F.A.; Alguacil, F.J. Carbon nanofibers: A new adsorbent for copper removal from wastewater. *Metals* **2018**, *8*, 914. [CrossRef]
40. Morales, A.E.; Mora, E.S.; Pal, U. Use of diffuse reflectance spectroscopy for optical characterization of un-supported nanostructures. *Rev. Mex. Fisica* **2007**, *53*, 18–22.
41. Güler, Ö.; Güler, S.H.; Başgöz, Ö.; Albayrak, M.G.; Yahia, I.S. Synthesis and characterization of ZnO-reinforced with grapheme nanolayer nanocomposites: Electrical conductivity and optical band gap analysis. *Mater. Res. Express* **2019**, *6*, 095602. [CrossRef]
42. Tien, H.N.; Luan, V.H.; Ho, L.T.; Kho, N.T.; Hahn, S.H.; Chung, J.S.; Shin, E.W.; Hur, S.H. One-pot synthesis of a reduced graphene oxide–zinc oxide sphere composite and its use as a visible light photocatalyst. *Chem. Eng. J.* **2013**, *229*, 126–133. [CrossRef]
43. Pandy, P.; Kurchania, R.; Haque, F.Z. Structural, diffuse reflectance and photoluminescence study of cerium doped ZnO nanoparticles synthesized through simple sol-gel method. *Optik* **2015**, *126*, 3310–3315. [CrossRef]
44. Durmus, Z.; Kurt, B.Z.; Durmus, A. Synthesis and characterization of Graphene Oxide/Zinc Oxide (GO/ZnO) nanocomposite and its utilization for photocatalytic degradation of basic fuchsin dye. *ChemistrySelect* **2019**, *4*, 271–278. [CrossRef]

 *nanomaterials*

Article

New Electrospun ZnO:MoO$_3$ Nanostructures: Preparation, Characterization and Photocatalytic Performance

Petronela Pascariu [1,*], Mihaela Homocianu [1], Niculae Olaru [1], Anton Airinei [1] and Octavian Ionescu [2,*]

[1] "Petru Poni" Institute of Macromolecular Chemistry, 41A Grigore Ghica Voda Alley, 700487 Iasi, Romania; mihaela.homocianu@icmpp.ro (M.H.); nicolae.olaru@icmpp.ro (N.O.); anton.airinei@icmpp.ro (A.A.)

[2] National Institute for Research and Development in Microtechnologies-IMT Bucharest, 126A, Erou Iancu 8 Nicolae Street, 077190 Bucharest, Romania

* Correspondence: dorneanu.petronela@icmpp.ro (P.P.); octavian.ionescu@imt.ro (O.I.)

Received: 30 June 2020; Accepted: 22 July 2020; Published: 28 July 2020

Abstract: New molybdenum trioxide-incorporated ZnO materials were prepared through the electrospinning method and then calcination at 500 °C, for 2 h. The obtained electrospun ZnO:MoO$_3$ hybrid materials were characterized by X-ray diffraction, scanning and transmission electron microscopies, ultraviolet (UV)-diffuse reflectance, UV–visible (UV–vis) absorption, and photoluminescence techniques. It was observed that the presence of MoO$_3$ as loading material in pure ZnO matrix induces a small blue shift in the absorption band maxima (from 382 to 371 nm) and the emission peaks are shifted to shorter wavelengths, as compared to pure ZnO. Also, a slight decrease in the optical band gap energy of ZnO:MoO$_3$ was registered after MoO3 incorporation. The photocatalytic performance of pure ZnO and ZnO:MoO$_3$ was assessed in the degradation of rhodamine B (RhB) dye with an initial concentration of 5 mg/L, under visible light irradiation. A doubling of the degradation efficiency of the ZnO:MoO$_3$ sample (3.26% of the atomic molar ratio of Mo/Zn) as compared to pure ZnO was obtained. The values of the reaction rate constants were found to be 0.0480 h^{-1} for ZnO, and 0.1072 h^{-1} for ZnO:MoO$_3$, respectively.

Keywords: molybdenum trioxide-incorporated ZnO; structural characterization; optical properties; photocatalytic activity

---

1. Introduction

Currently, many efforts are being made worldwide to develop new high-performance photocatalytic materials for energy and environmental applications. Metal oxide semiconductor materials in various shapes and structures, including ZnO, TiO$_2$, CuO, and MgO, have proven to be a good alternative for the degradation of various organic dyes. It is known that ZnO is an oxide semiconductor having a broad direct band gap (3.37 eV), high excitation binding energy (60 meV), and good electrical, mechanical, and optical and photocatalytic properties, comparable to those of TiO$_2$. Metal doping of ZnO can significantly improve the photocatalytic activity, it is thought by (i) generation of trapping site which will decrease the recombination rate of photoinduced electron-hole pairs; (ii) decrease of band gap energy of photocatalysts; and (iii) structural defects in the crystalline phase of ZnO. Many studies have been based on the development of new photocatalysts based on ZnO doped with different metals (Ag, La, Er, Sm, Cu, Au, Ce, Ni, Fe, etc.) to improve the photocatalytic activity for the degradation of different organic dyes and to extend the degradation domain using visible light [1,2]. For example, Pascariu et al. reported good photocatalytic responses of ZnO-SnO$_2$ nanostructures used for rhodamine B dye degradation with an initial dye concentration of 0.01 mM and a catalyst dosage of 0.5 g/L [3]. Likewise, the same authors obtained an improvement of the photocatalytic activity after doping ZnO with Ni or Co [4]. Molybdenum trioxide (MoO_3) is a very interesting transition metal oxide having a wide band gap energy of about 3 eV, distinctive optical properties, and highly visible-light photocatalytic activity [5,6]. But the synthesis of $ZnO-MoO_3$ nanostructures (especially by the electrospinning method) and their detailed properties after being incorporated into the ZnO matrix are less reported compared to other semiconductor oxides, such as SnO_2, ZnO, TiO_2 and In_2O_3 [7,8]. Moreover, it is known that the properties of molybdenum trioxide are strongly dependent on the synthesis methods and can be greatly modulated by doping it with other metal oxides [9]. There are many methods to develop new materials of different shapes and sizes, including hydrothermal, sol-gel, precipitation, microemulsion, solvothermal, the electrochemical deposition process, microwave, polyol, wet chemical method, flux methods and electrospinning [10]. The electrospinning method is one of the simplest, cheapest, and most efficient for obtaining nanostructured materials. Also, this method offers the possibility of obtaining materials with controllable diameters of the fibers, very high surface-to-volume ratio, and specific surface and excellent functional properties [1]. Also, the electrospinning technique is intensively used in obtaining ceramic materials. One study based on Mo-doped ZnO materials obtained by the electrospinning method has been reported by Kim et al. [11], and their investigation regarding gas-sensing properties in ethanol, trimethylamine (TMA), CO and H_2 medium. To the best of our knowledge, the development of $ZnO:MoO_3$ nanostructures by the electrospinning-calcination method and then their testing for dye degradation have not been reported in the literature so far. Similar systems have been obtained by other methods, such as coating of MoO_3 altered ZnO, by surface metal impregnation [12], $ZnO@MoO_3$ core/shell nanocables by the electrodeposited method [13], and 1D/1D $ZnO@h-MoO_3$ synthesized via the solid state impregnation-calcination method [14].

Therefore, in this paper, we proposed the preparation of new pure ZnO and $ZnO:MoO_3$ nanostructures by the electrospinning method and then calcination at 500 °C for 2 h. The polymer solutions were prepared from polyvinyl alcohol (PVA) dissolved in water. The details of the structural and optical properties of the obtained MoO_3 incorporated-ZnO nanostructures were comparatively analyzed and discussed. Moreover, the photocatalytic performance of these materials was evaluated by photodegradation of RhB dye in aqueous solution under simulated sunlight irradiation.

2. Materials and Methods

2.1. Materials

Polyvinyl alcohol (PVA) (Mn = 80.000), zinc acetate [$Zn(CH_3COO)_2 \cdot 2H_2O$], and ammonium molybdate ((NH_4)$_6Mo_7O_{24} \cdot 4H_2O$) were purchased from Sigma-Aldrich (Merck KGaA, Darmstadt, Germany). Deionized water was used as a solvent.

2.2. Preparation of ZnO:MoO₃ Nanostructures

We dissolved 0.55 g PVA powder in 5 mL of deionized water and then heated at 100 °C under vigorous magnetic stirring for 8 h and thus the solution for electrospinning was obtained. This solution was cooled to room temperature and then 0.4 g of zinc acetate was added to it. Finally, ammonium molybdate in various concentrations was added to the above electrospun solution. This prepared viscous solution was stirred for another 3 h and then transferred into a needle syringe with a diameter of 0.5 mm. All new nanostructures were obtained using a home-made electrospinning device [15,16]. The optimal conditions for the production of these fibers before calcination were as follows: high voltage source (25 kV), the distance between the needle tip and the collector (stainless steel foil) was 15 cm at a flow-rate of about 0.75 mL/h. The $ZnO:MoO_3$ nanostructures were obtained after calcination in an oven (in air) at 500 °C for 2 h. In this study, a series of four materials (with different concentrations of MoO_3) were prepared and are presented. More details regarding growth and sample names correlation with the specific growth conditions can be found in Table 1.

Table 1. Results of EDX (energy-dispersive X-ray spectroscopy) measurements for the pure ZnO, MoO_3, and ZnO-incorporated MoO_3 nanostructures.

Sample Codes	Composition of Precursor Salt	Atomic Molar Ratio Mo/Zn (%)	Atomic Concentration (%) Obtained from EDX		
			Zn (%)	O (%)	Mo (%)
S1	0.4 g $Zn(NO_3)_2 \cdot 6H_2O$	-	53.08	46.92	-
S2	0.4 g $(NH_4)_6Mo_7O_{24} \cdot 4H_2O$	-	-	72.23	26.77
S3	0.4 g $Zn(NO_3)_2 \cdot 6H_2O$ 0.002 g $(NH_4)_6Mo_7O_{24} \cdot 4H_2O$	0.84	51.20	47.85	0.92
S4	0.4 g $Zn(NO_3)_2 \cdot 6H_2O$ 0.004 g $(NH_4)_6Mo_7O_{24} \cdot 4H_2O$	1.66	53.86	44.62	1.51
S5	0.4 g $Zn(NO_3)_2 \cdot 6H_2O$ 0.008 g $(NH_4)_6Mo_7O_{24} \cdot 4H_2O$	3.26	53.29	44.67	2.30
S6	0.4 g $Zn(NO_3)_2 \cdot 6H_2O$ 0.012 g $(NH_4)_6Mo_7O_{24} \cdot 4H_2O$	4.81	50.55	44.62	4.82

2.3. Structural Characterization

X-ray diffraction (XRD) patterns of $ZnO:MoO_3$ nanostructures were recorded using a Rigaku SmartLab-9kW X-ray diffractometer (Rigaku Corporation, Japan). SEM (scanning electron microscopy)/energy-dispersive X-ray spectroscopy (EDX) measurements were performed using a JEOL JSM 6362LV (Japan) electron microscope coupled with an EDAX INCA X-act Oxford Instrument detector (Oxford, UK). Transmission electron microscopy (TEM) studies were performed using a Hitachi HT7700 Dual Mode STEM (Japan) in TEM mode. To perform the TEM studies, a small quantity of material was fixed onto Cu mesh grids after dispersion in ethanol and sonication.

2.4. Steady-State Spectral Measurements

The optical measurements of the samples were evaluated by ultraviolet–visible (UV–vis) reflectance spectra recorded with a SPECORD 210Plus Analytik Jena (Jena, Germany) spectrophotometer equipped with an integrating sphere. The band gap energy values were found from the optical data solution. Steady state absorption and fluorescence spectra were recorded with a spectrophotometer SPECORD 210Plus Analytik Jena (Jena, Germany) and on an Edinburgh FLS980 spectrometer (Edinburgh, UK), respectively.

2.5. Time-Resolved Fluorescence

The time-resolved fluorescence spectra were collected on an Edinburgh FLS980 spectrometer using a time correlated single photon counting method. The emission decay profiles for all nanostructures dispersed in 1-propanol solution were determined in a 10×10 mm quartz cell, excited by a nanosecond diode laser (EPL-375) (Edinburgh Instruments Ltd., Livingston, UK) operating at 375 nm as light source. The fluorescence decays were evaluated using the nF9000 software attached to the equipment (Edinburgh Instruments Ltd., Livingston, UK) and the best fitted parameters were obtained for the reduced chi-squared values close to 1, and the weighted residuals were uniformly distributed around the zero line.

2.6. Nanosecond Transient Absorption Spectroscopy

Nanosecond transient absorption experiments were performed using a nanosecond laser flash photolysis technique (LP980, Edinburgh Instruments, (Livingston, UK)). The spectrometer was connected to a laser source (Ekspla NT342), which allows us to generate a high concentration of excited state species, laser pulses at 372 nm, and a frequency of 1 Hz. The absorption lifetime values of excited state species were evaluated with the L900 software attached to the equipment ((Edinburgh Instruments Ltd., Livingston, UK).

2.7. Photocatalysis Tests

The photocatalytic activity of the MoO_3-incorporated ZnO nanostructures was tested by photodegradation of rhodamine B (RhB) dye in aqueous solution under visible light irradiation using the same experimental degradation procedure as reported in previous work [3]. Briefly, 5 mg of the catalyst was dispersed in a vial containing 10 mL of RhB dye solution (5 mg/L), having a controlled temperature at 25 °C. A 100 W tungsten lamp was served as the light source. The power of light source was 102.74 $kJ \cdot m^{-2} \cdot h^{-1}$, and measured by a PMA 2100 apparatus, prod by Solar Light Co. (Glenside, PA, USA). The wavelength range of the tungsten bulb varies between 350 and 850 nm. The initial concentrations of RhB dye and after irradiation at different times were determined using a UV–vis spectrometer SPECORD 210Plus, Analytik Jena (Jena, Germany).

3. Results and Discussion

3.1. Morphological Characterization

Some examples of SEM micrographs of $ZnO:MoO_3$ materials obtained after calcination at 500 °C for 2 h are shown in Figure 1. Figure 1a,b depict the morphology of pure ZnO and MoO_3 metal oxides. The as-grown ZnO exhibits a cylindrical microrods structure with a rough surface. The SEM image (Figure 1b) corresponding to the MoO_3 sample reveals a morphology composed of microparticles, with a platelet structure and diameters between 4–5 μm. This different structure of the two types of materials (ZnO and MoO_3) leads to a morphology composed of different types of crystals interconnected between them in the $ZnO:MoO_3$ nanostructures. TEM studies were performed to confirm the intimate structure of the $ZnO:MoO_3$ materials. TEM observation proved the formation of two kinds of well-shaped nanosized crystallites: one with dimensions of about 20–30 nm and the second, with larger size of ~40–45 nm. An example of TEM micrographs of nanocrystalline building blocks on sample S5 is presented in Figure 2.

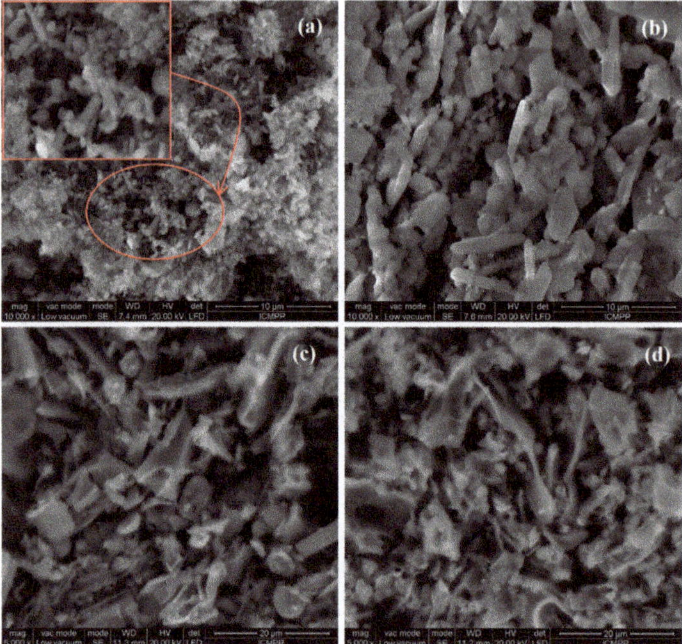

Figure 1. Scanning electron microscope (SEM) images of the ZnO (S1) (**a**), MoO_3 (S2) (**b**), $ZnO:MoO_3$ (S4) (**c**) and $ZnO:MoO_3$ (S5) (**d**) nanostructures.

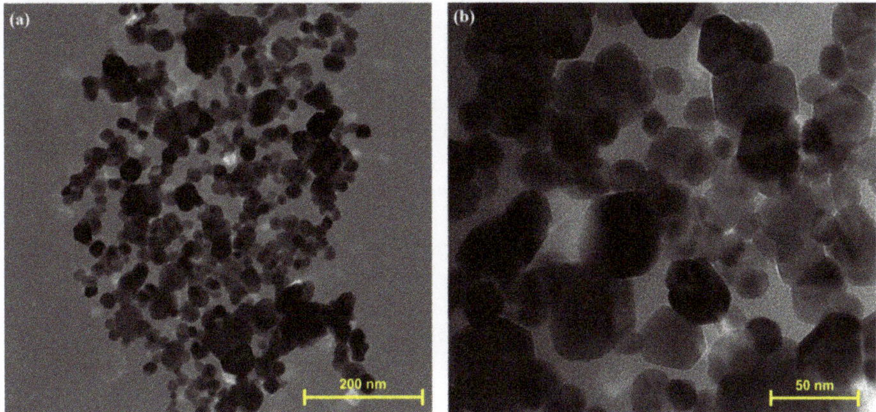

Figure 2. Transmission electron microscope (TEM) micrographs of nanocrystalline building blocks on sample (**a**) S5 × 350 k; (**b**) × 1300 k.

EDX analysis was used to report the elemental composition of all samples. The EDX spectra (Figure 3) confirm the presence of Zn and O in a ratio of about 1:1, corresponding to the ZnO sample and the presence of Mo and O in a ratio of 1:3 corresponding to the MoO_3 sample. The atomic percentages of Mo in composite nanostructures were: 0.92% (S3), 1.51% (S4), 2.30% (S5), and 4.82% (S6), respectively (Table 1). Also, the EDX measurements have predicted the expected values for Mo according to the prepared samples.

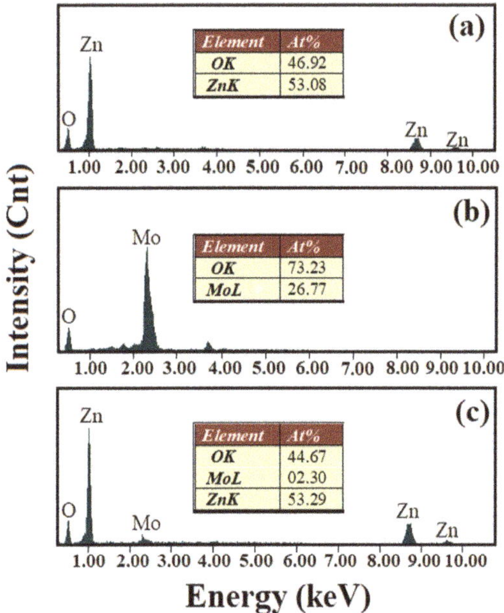

Figure 3. EDX spectra of pure ZnO (**a**), MoO3 (**b**) and ZnO:MoO3 (S5) (**c**) nanostructures.

3.2. X-ray Diffraction (XRD) Analysis

The crystalline structure of pure ZnO, MoO_3 and ZnO:MoO$_3$ composites were confirmed using XRD measurements, and the XRD patterns are illustrated in Figure 4. The diffraction peaks of ZnO are indexed as hexagonal wurtzite-type structure (JCPDS No. 00-230-0112, space group: P63mc, No. 186) having as main peaks of diffractions at 2θ = 31.84°, 34.51°, 36.32°, 47.62°, 56.66°, 62.92°, 66.43°, 67.98° and 69.13°, attributed to diffraction planes (110), (002), (101), (102), (110), (103), (200), (112), (201), respectively. The diffraction pattern of α-MoO_3 confirms the orthorhombic crystalline structure Joint Committee on Powder Diffraction Standards (JCPDS No. 00-900-9670, space group: Pbnm, No. 62) with the main diffraction peaks at 23.36°, 25.65°, 27.23°, 29.59°, indexed by (110), (040), (021) and (130). The XRD data corresponding to the $ZnO{:}MoO_3$ composites show a mixed crystalline structure formed from ZnO with hexagonal wurtzite structure and MoO_3 with orthorhombic structure. The average crystallite size was calculated according to the Scherrer relation $D = 0.8\lambda/\beta cos\theta$, were λ is the X-ray wavelength corresponding to CuKα radiation, β is the full width at half maximum of the peak and θ is the Bragg angle. The average crystallite size was found to be D = 21.77 nm for pure ZnO and D = 14.93 nm is corresponding to $ZnO{:}MoO_3$ nanostructures, respectively. The presence of reflections corresponding to ZnO and MoO_3 in ZnO-MoO_3 nanostructures shows the successful formation of the nanocomposite. This decrease in the crystallite size of $ZnO{:}MoO_3$ (S5) may be attributed to the formation of Mo–O–Zn bands on the surface of the doped materials, which inhibits the growth of the crystallite, as reported by many authors for similar systems [17,18]. The decrease in the crystallite size of $ZnO{:}MoO_3$ nanostructures can be determined by the lattice distortion by Mo incorporation due to the difference between the ionic radius of Mo^{6+} (0.065 nm) smaller them that of Zn^{2+} (0.074 nm). The lattice constants a and b were 3.252 Å, 5.209 Å for ZnO and 3.249 Å, 5.207 Å obtained for $ZnO{:}MoO_3$ (S5), values that are in good agreement with the standard ones (a = 3.253 Å, c = 5.213 Å for ZnO → JCPDS 34-1451) [19].

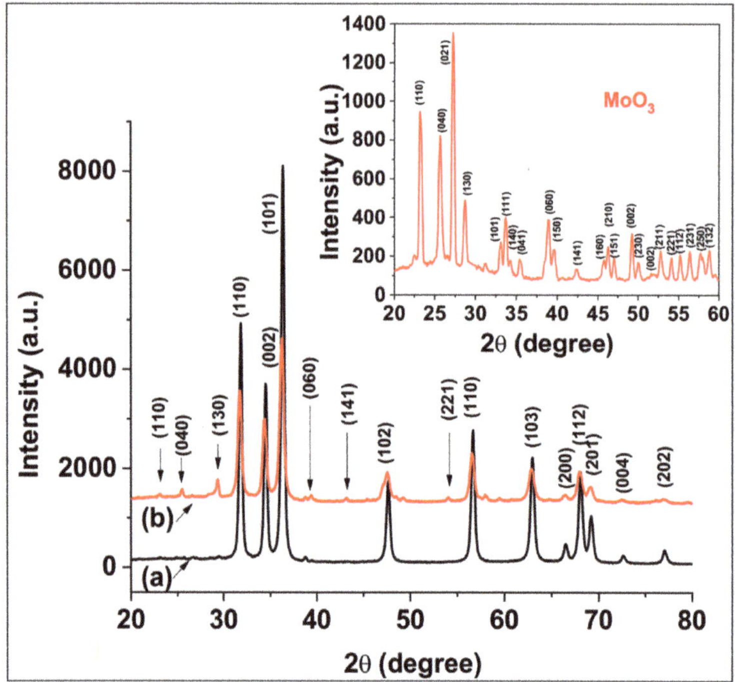

Figure 4. X-ray diffraction (XRD) patterns for pure ZnO (**a**), MoO_3 (**inset**) and $ZnO{:}MoO_3$ (S5) (**b**) composite material.

3.3. Optical Properties

Evaluation of optical properties of materials is a very important factor for the study of their photocatalytic activity. UV–vis absorption and emission spectra of MoO_3-incorporated ZnO nanostructures dispersed in 1-propanol solution were investigated at room temperature. The incorporation of MoO_3 into the ZnO matrix modify the optical properties of the obtained nanostructures. Figure 5. showed the absorption spectra of the pure ZnO- and MoO_3-incorporated ZnO samples with various MoO_3 weight percentages (S1 → S5). It can be observed that the absorption of all the MoO_3-incorporated ZnO samples did not show a large difference in the UV region (only small variations (1–2 nm) in the absorption band position in the spectra). In contrast, the absorption maxima of all MoO_3-incorporated ZnO samples shift towards blue (from 382 to 371 nm) are observed, compared with the absorption maximum of the pure ZnO samples. These shifts are due to the incorporation of the MoO_3 materials in the pure ZnO matrix. From Figure 5, it can be remarked that the absorption of ZnO:MoO_3 nanocomposites was enhanced when the MoO_3 doping level did not exceed 1.66% molar ratio Mo/Zn, while for nanostructures with a higher content of MoO_3 the absorption shows a small decrease (Figure 5).

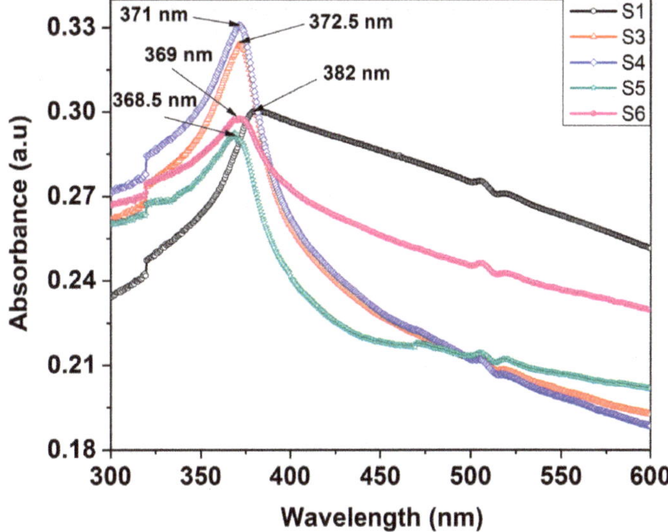

Figure 5. Absorption spectra of pure ZnO and ZnO-incorporated MoO_3 nanostructures with different atomic weight percentages of MoO_3 oxide.

One of the important effects of the MoO_3 oxide incorporation into the ZnO matrix is the reduction of the direct band gap energy value from 3.211 eV (corresponding to ZnO) to 3.170 eV for ZnO:MoO_3 composite materials (Table 2). The band gap energy values of the pure ZnO, MoO_3, and for the ZnO:MoO_3 composite materials were determined by UV–vis measurements using reflectance spectra, registered between 300–600 nm. Figure 6a showed the reflectance spectra corresponding to pure materials (ZnO, MoO_3) and composite systems ZnO:MoO_3 (S3 → S6). Thus, the optical band gap energy values, E_g, for all materials were calculated using the Kubelka–Munk equation presented below:

$$F(R) = \frac{(1-R)^2}{2R} \tag{1}$$

where $F(R)$ is the Kubelka–Munk function and R is the diffuse reflectance of materials, respectively. Additionally, using $[F(R)h\vartheta] = c\left(h\vartheta - E_g\right)^n$ equation: where $h\vartheta$ is photon energy, n is a constant giving the type of optical transition, c is a constant and E_g denotes the band gap energy.

Table 2. Calculated band gap energies and band edge potentials for pure ZnO and ZnO-incorporated MoO_3 nanostructures.

Codes	Eg	E_{VB} (eV)	E_{CB} (eV)
S1	3.210	2.890	−0.310
S3	3.186	2.883	−0.303
S4	3.177	2.878	−0.298
S5	3.170	2.875	−0.295
S6	3.189	2.884	−0.304

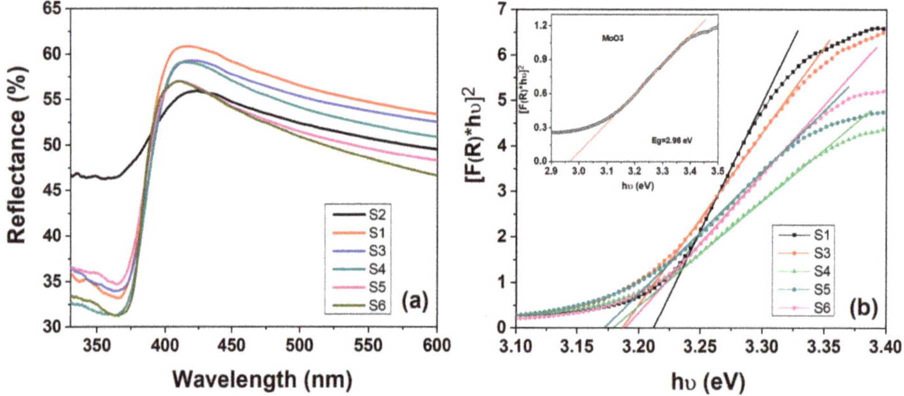

Figure 6. Diffuse reflectance ultraviolet–visible (UV–vis) spectra (**a**) and plots of $[F(R)h\vartheta]^2$ versus $h\upsilon$ (**b**) for pure ZnO (S1), MoO_3 (S2) (**inset**), and ZnO:MoO_3 (S3 → S6) nanostructures.

The plots of $[F(R)h\vartheta]^2$ as a function of the photon energy ($h\vartheta$) of all the materials are presented in Figure 6b and by extrapolating the linear part of each curve, the values of the direct optical band gap (Eg) were obtained. The pure ZnO material has an Eg value of 3.21 eV, this value is in good agreement with the reported literature for ZnO [20]. Instead, for pure MoO_3 material, a lower Eg value of 2.96 eV (closer to the literature data) [9] was obtained. The band gap energies of the composite materials based on ZnO:MoO_3 were calculated to be: 3.186 eV (S3), 3.177 eV (S4), 3.170 eV (S5), and 3.189 eV (S6), respectively. The presence of MoO_3 dopant in the ZnO matrix induces a small shift of the absorption edge of ZnO (see Figure 6) in the long wavelength direction and thus decreases the band gap energy (Eg) values. Based on the values of Eg one can calculate the positions of the valence band (VB) and the conduction band (CB) (redox abilities) of new ZnO:MoO_3 nanostructures, which directly influence their photocatalytic activity. The conduction band (CB) and valence band (VB) potentials of ZnO:MoO_3 were calculated using the following equations [21].

$$E_{VB} = \chi - E_e + 0.5E_g \tag{2}$$

$$E_{CB} = E_{VB} - E_g \tag{3}$$

where χ and E_e, are the electronegativity (for ZnO, the value is approximately 5.79 eV) [22] and the free electron energy on the hydrogen scale (approximately 4.5 eV), respectively. The calculated values of E_{VB} and E_{CB} are listed in Table 2. The values of E_{CB} for MoO_3-incorporated ZnO nanostructures shifted to lower energy compared to the pure ZnO, which caused a narrower band gap.

To explore the charge carrier trapping, migration, and charge transfer transitions in the ZnO:MoO$_3$ hybrid nanostructures, the emission spectra of these new materials were recorded under 375 nm wavelength excitation (see Figure 7).

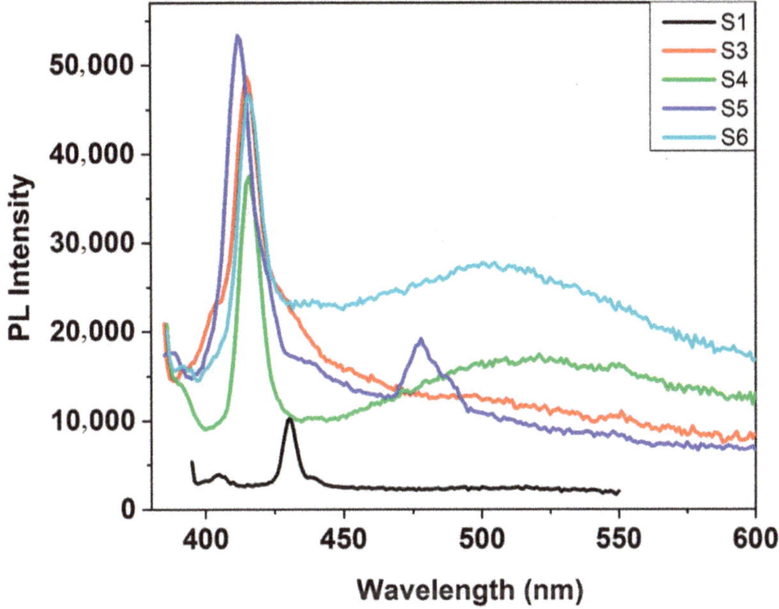

Figure 7. Fluorescence spectra of pure ZnO and ZnO:MoO$_3$ nanostructures, recorded in 1-propanol solution, at room temperature.

The emission spectra of MoO$_3$-incorporated ZnO nanostructures show a sharp and intense blue emission band around 412–416 nm, ascribed to the near band edge (NBE) emission due to free exciton recombination [23] and a broader and lower green emission band in the visible region at $\approx$ 450–550 nm (with maxima around 506 nm). This green emission band is due to the formation of oxygen vacancies by the presence of MoO$_3$ oxide in the structure of the target nanostructures and the presence of some intrinsic defects in the ZnO structure. Generally, the intrinsic defects in ZnO include oxygen vacancies (V$_o$), zinc vacancies (V$_{Zn}$) oxygen (O$_i$), zinc (Zn$_i$) interstitials, and defect states dominating the emission in the visible range [24]. It was observed from Figure 7 that the emission intensity is considerably enhanced with the addition of MoO$_3$ as compared to the free ZnO sample. This remarkable improvement in the emission of MoO$_3$ incorporated samples can be attributed to the energy transfer from ZnO to the MoO$_3$ metal oxide and a higher recombination rate in these samples. Instead, in the photoluminescence spectrum of pure ZnO nanostructures, the wavelength of the maxima of the emission band is located at 431 nm, with low intensity, and the green emission band has disappeared (compared to the spectra of MoO$_3$-incorporated ZnO samples). The undoped zinc oxide nanostructure displays only an emission band around 431 nm, with low intensity and the green emission band practically disappeared in comparison with MoO$_3$-incorporated ZnO samples. Also, it can be seen that the intensity of the green emission band increases as the molybdenum trioxide level increases, excepting nanostructure ZnO:MoO$_3$ (S5), where an emission band at about 478 nm has occurred (Figure 7), which have not been observed in the other samples due to the intense transitions determined by the oxygen vacancies. Moreover, the lower intensities of the green emission band in sample S5 can determine the creation of more electron trapping sites, which facilitates the transfer of photogenerated charge carriers increasing thus the electron-hole lifetime and the photocatalytic response was exchanged [25]. As seen from Table 3, the nanostructure ZnO:MoO$_3$ (S5) has a larger decay time as compared to other samples. In fact, for this nanostructure the highest shift to shorter wavelengths of the absorption and emission bands was observed and at the same time, the lowest value of E$_g$ was obtained (Table 2). The weak blue emission band at 478 nm can be assigned to the electron transitions between interstitial zinc (Zn$_i$) and Zn vacancies [26]. Thus, the sample ZnO:MoO$_3$ (S5) practically exhibits a very weak green emission band because the MoO$_3$ loading can increase distortion centers and some surface defects in the lattice leading to the decrease of the oxygen vacancies. Furthermore, the emission spectra of zinc oxide (ZnO) sample incorporating with molybdenum trioxide (MoO$_3$) appear similar in shape to the emission peaks but show a small shift (19 nm) in wavelength toward the blue range as compared to that observed for pure ZnO nanostructures.

3.4. Time-Resolved Photoluminescence Data and International Commission on Illumination (CIE) Coordinates

For this purpose, the samples (S1 → S6) were exciting with a light source of 375 nm. Base on the maximum emission band at 406 nm, the lifetime decay curves were evaluated. As a representative example, in Figure 8 is displayed the lifetime decay profiles for the ZnO:MoO$_3$ (S5) sample. The decay curves were fitted to a bi-exponential equation for all the studied ZnO:MoO$_3$ nanostructures and the experimental emission lifetime values of these materials were calculated using the equations given below [23]:

$$I(t) = A_1 \exp\left(-\frac{t}{\tau_1}\right) + A_2 exp\left(-\frac{t}{\tau_2}\right) \tag{4}$$

and the value of average decay time can be estimated by the following formula:

$$\tau^*_{eff}\ (ns) = (A_1\tau_1^2 + A_2\tau_2^2)/(A_1\tau_1 + A_2\tau_2) \tag{5}$$

where $I(t)$ is the time dependent emission intensity (emission intensity at any time); A_1 and A_2 are the fitting constants and have values between [0, 1]; τ_1 and τ_2 are the decay times of the fast and slow decay components.

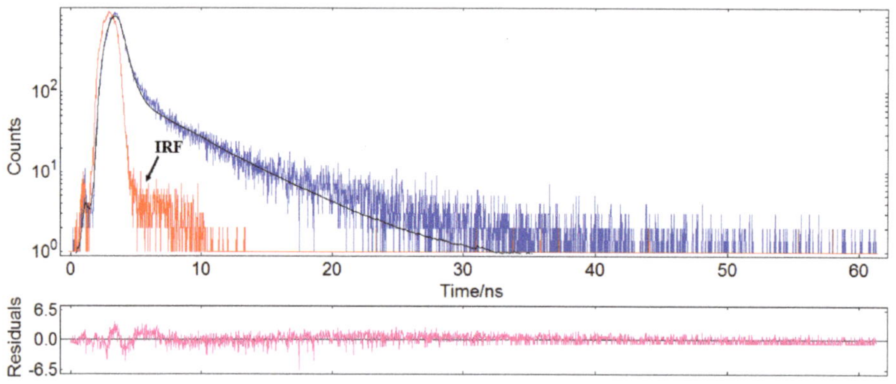

Figure 8. The lifetime decay profiles for the ZnO:MoO$_3$ (S5) sample (λ_{ex} = 406 nm); IRF, instrument response function.

The obtained values of the fast and slow decay components, as well as average lifetime values for all samples, are listed in Table 3. In general, the fast decay component is related to the non-radiative recombination and the slow decay is connected with the radiative lifetime of the free exciton. It was stated that the non-radiative recombination process is featured in defects related to oxygen vacancies [27]. The emission decay for all investigated nanostructures was obtained by using a biexponential model to fit the experimental data and have resulted in two lifetimes τ_1 and τ_2. For all investigated nanostructures

τ_1 is lower than τ_2, but the contributions of the two lifetime components (a_1 and a_2) are opposite (Table 3). Therefore, the highest lifetime τ_1 was obtained for the ZnO:MoO$_3$ (S5) sample (0.465 ns). Both fast and slow decay constants can be tuned by adding MoO$_3$ to the ZnO matrix. The irregular variations of lifetime values as a function of the MoO$_3$ content can be explained by the difference of dispersion of MoO$_3$ in these samples, where the redundant MoO$_3$ can act as recombination centers [28]. Since MoO$_3$ has a lower band gap compared to ZnO will induce a shorter lifetime for the generated electron-hole pairs, decreasing the efficiency. For the pure ZnO sample, the average lifetime values are shorter than that of all ZnO:MoO$_3$ nanostructures (Table 3). This increase in the average lifetime values may be attributed to the MoO$_3$ doping-induced non-radiative recombination centers. Moreover, the average lifetime values (Table 3) decrease with increasing doping concentration (from S1 to S5), which indicates the existence of an energy transfer process between the two components.

Table 3. Fitted decay times and average lifetimes (τ^*_{eff} (ns)) for all pure ZnO and ZnO:MoO$_3$ nanostructures.

Codes	Fast Decay Component		Slow Decay Component		χ^2	τ^*_{eff} (ns)
	τ_1 (ns)	a_1 (%)	τ_2 (ns)	a_2 (%)		
S1	0.0664	74.06	4.1058	25.94	1.004	3.93
S3	0.1068	51.95	5.3416	48.05	1.034	5.23
S4	0.0624	76.76	4.6088	23.24	0.995	4.42
S5	0.4654	72.68	5.0603	27.32	1.007	4.17
S6	0.0662	74.05	4.0695	25.95	1.086	3.89

Based on the emission spectra of the pure ZnO and MoO$_3$-incorporated ZnO nanostructures was built their CIE (International Commission on Illumination)-1931 color chromaticity diagram, which is shown in Figure 9. The values obtained for the color CIE chromaticity coordinates for pure ZnO and all the hybrid composites are shown in Figure 9. Generally, the quality of any light emitted can be investigated in terms of correlated color temperature (CCT) which can be calculated using the McCamy's relation as [29]:

$$CCT = 437n^3 + 3601n^2 + 6861n + 5517 \tag{6}$$

where, n = (x − xe)/(ye − y) is the inverse slope line and the point at xe = 0.332, ye = 0.186 is the epicenter. The values obtained for calculated CCT were found to be: 82,868 K, 11,389 K, 187,053 K, and 16,098 K, for S3, S5, S5, and S6 nanostructures, respectively. These values are located inside of the range of clear blue poleward sky light [30]. However, based on the chromaticity diagram, it can be said that the samples exhibit some differences in color emission determined by the MoO$_3$ loading. As seen, the nanostructures containing 0.84% and 3.26% molar ratio Mo/Zn presented emissions in the blue region, whereas the samples having 1.66% and 4.81% molar ratio Mo/Zn exhibited coordinates close greenish-blue color. This color shift to the green region can be connected to a higher level of defects due to the high intensity of the green emission band around 520 nm, leading to a higher electron-hole pair recombination rate, which is not favorable to the photocatalysis process.

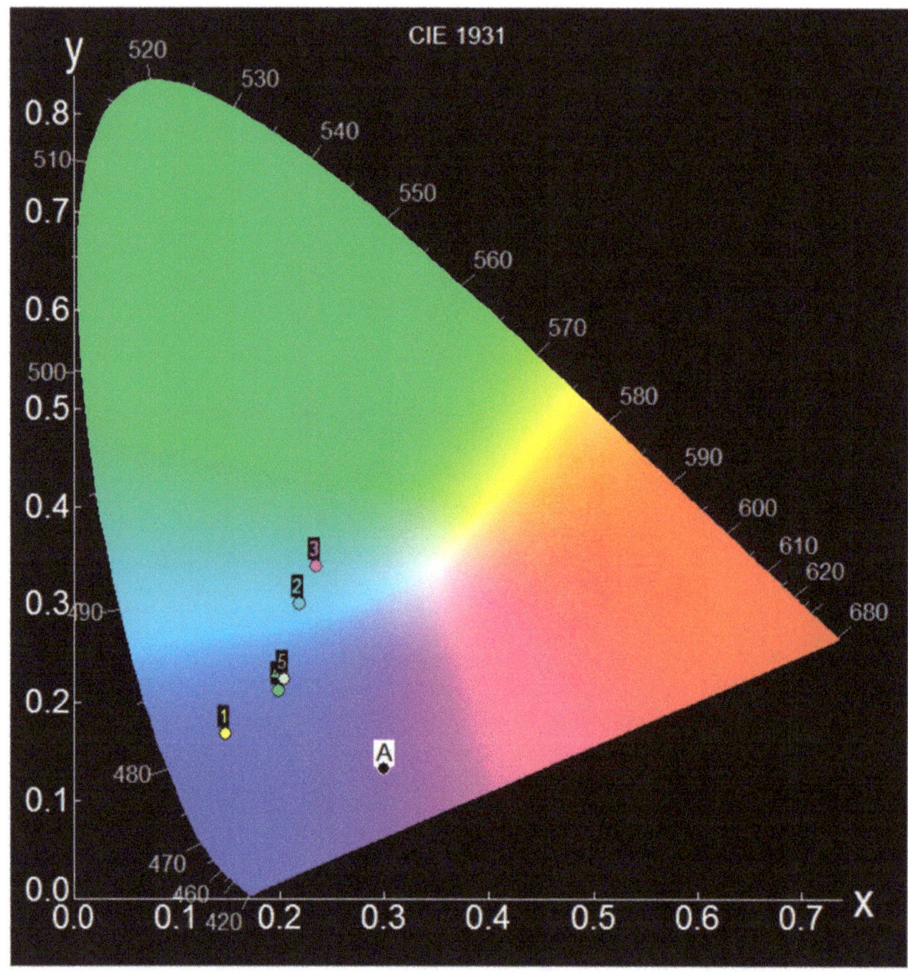

ID	Name	X	Y	Z
A	Ref	0.30000	0.13333	"0.00000×10^{0}"
1	S1	0.14858	0.16759	"1.93200×10^{2}"
2	S6	0.21912	0.30105	"3.68900×10^{3}"
3	S4	0.23558	0.33862	"2.44100×10^{3}"
4	S5	0.19864	0.21232	"1.41500×10^{3}"
5	S3	0.20479	0.22442	"1.67100×10^{3}"

Figure 9. The CIE color chromaticity diagram of pure ZnO and MoO$_3$-incorporated ZnO nanostructures.

3.5. Nanosecond Time-Resolved Absorption Data

For two selected samples, (S3 and S5), the transient absorption data in 1-propanol solution were recorded at room temperature in order to monitor the optical absorption of photogenerated transient species (such as excited states, radicals and solvated electrons produced by the interaction between radiation and the substrate, more specifically, the kinetics of each transient species). Figure 10 shows the kinetics data, for S3 (Figure 10a,b) and S5 (Figure 10c,d) samples, which were fitted to exponential decay using Edinburgh Instruments software, after recording the decay rates of the transient species monitored at 372 nm (the wavelength at which these transient species absorb, see Figure 5). The decay rate for the S5 sample was found to be faster than the decay rate obtained for the sample having S3 (107.44 ns).

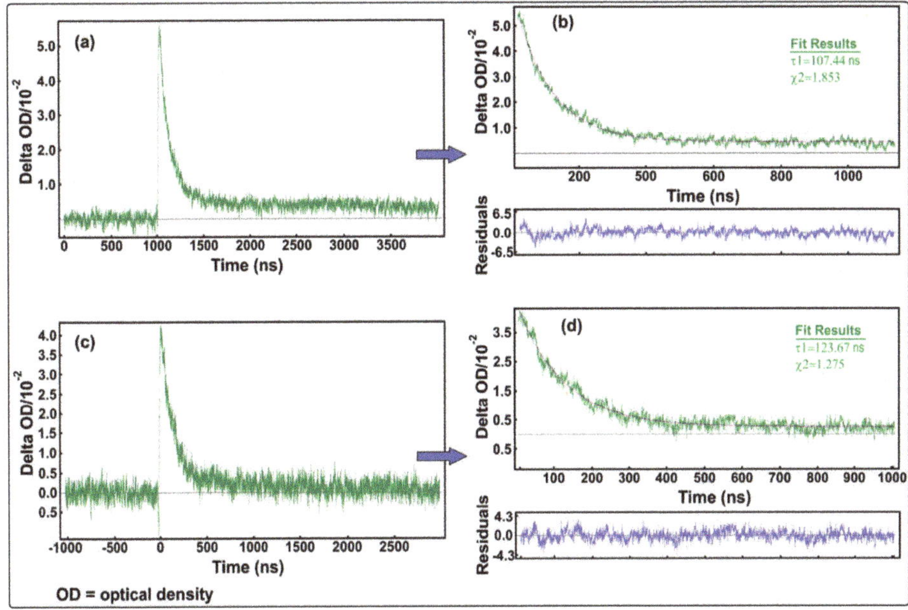

Figure 10. Kinetics transient absorption decay and exponential fitting for ZnO:MoO$_3$ (S3) (**a,b**) and ZnO:MoO$_3$ (S5) (**c,d**) samples, upon excitation at 372 nm.

3.6. Photocatalytic Properties

The photocatalytic activity of the pure ZnO and synthesized ZnO:MoO$_3$ nanocomposites was evaluated by photodegradation of rhodamine B (RhB) dye in aqueous solution, with an initial concentration of 5 mg/L, under visible light irradiation from a tungsten lamp (100 W)), at different times. The absorption spectra of the RhB dye (blank test) were analyzed at different exposure times (up to 20 h of irradiation), in the absence of the photocatalyst. The blank test showed that the intensity of the absorption bands of RhB decreases slightly reaching a degradation of about 5.8%. To get the most reliable results, the RhB dye degradation processes were analyzed in the presence of pure ZnO and compared with the ZnO:MoO$_3$ (S3 → S6) hybrid materials. Figure 11 shows changes in the absorption spectra of RhB dye during irradiation under visible light in the presence of ZnO (Figure 11a) and ZnO:MoO$_3$ (S5) (Figure 11b) nanostructures.

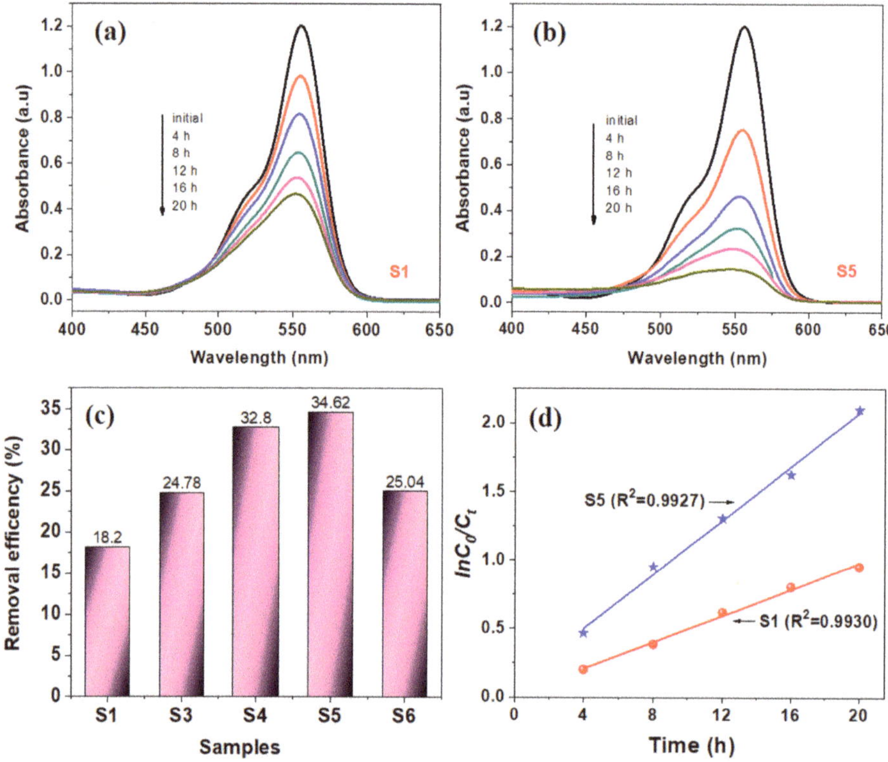

Figure 11. The absorption spectra changes of the RhB dye in aqueous solution, at various times, upon exposure to visible light irradiation in the presence of pure ZnO (S1) (**a**) and ZnO:MoO3 (S5) (**b**) samples; Removal efficiency (%) of the degradation of RhB dye, after 6 h of visible light irradiation (**c**); Pseudo-first order kinetics for RhB dye degradation in the presence of S1 and S5 samples, respectively (**d**).

The intensity of the absorption bands in the absorption spectra of the RhB dye solution shown in Figure 11a,b decreases with increasing illumination time. It can be observed that the intensity of the absorption band at 556 nm decreased with increasing irradiation time, but with a maximum decrease observed for the S5 sample (RhB dye was highly degraded in this case). The removal efficiency (RE(%)) was calculated based on the following equation:

$$RE(\%) = \frac{(C_0 - C_e)}{C_0} \times 100\%$$

(7)

(where C_0 is the initial RhB concentration (mg/L) and C_e is the RhB concentration at the time t (mg/L)) and the results obtained after the degradation of RhB dye for 6 h of irradiation in the presence of pure ZnO and ZnO:MoO$_3$ are presented in Figure 11c. The removal efficiency of all samples obeys the followed order of S5 > S4 > S3 > S6 > pure ZnO (S1), that was very close to fluorescence data (fluorescence intensity). The higher photocatalytic efficiency of the ZnO:MoO$_3$ (S5) nanostructures can be accounted for by decreasing Eg values compared to pure ZnO, a fact confirmed by other authors in differently doped ZnO-based systems [31,32]. Moreover, the increase in photocatalytic activity with the content of MoO$_3$ may be due to the formation of MoO$_3$/ZnO heterojunctions, as reported by Hirotaka et al. [33] for a similar system (TiO$_2$/MoO$_3$).

To estimate quantitatively the kinetics of RhB dye degradation, we used a pseudo-first order model expressed by the following equation: $\ln(C_0/C) = kt$, (where, C_0 and C are the concentrations of dye in the solution at time 0 and t, respectively, and k is the pseudo-first-order rate constant). From the linear plots of $\ln(C/C_0)$ versus the irradiation time (see Figure 11d), having a good correlation with the pseudo-first order reaction kinetics ($R^2 > 0.99$), the values of the reaction rate constants were calculated and were found to be: $k_1 = 0.0480$ h^{-1} (ZnO) and $k_2 = 0.1072$ h^{-1} (ZnO:MoO$_3$ (S5)), respectively. Likewise, these results show that after doping with 3.26% molar ratio Mo/Zn, the value of the reaction rate constant increases significantly, doubling its value ($k_2 = 2 \times k_1$), as compared to that obtained for pure ZnO. The efficiency of photocatalytic degradation is found to be better in the presence of the MoO$_3$-incorporated ZnO nanostructures than those of pure ZnO and/or MoO$_3$ nanostructures [17], due to the reduction in band gap energy. The presence of structural defects due to the oxygen vacancies (V_O), Zn-vacancy (V_{Zn}), oxygen interstitial (O_i), Zn-interstitial (Zn_i) and the extrinsic impurities, which have been confirmed by photoluminescence measurements, also leads to an increase of photocatalytic efficiency for ZnO:MoO$_3$ (S5) sample. Similar behavior has been also reported by other authors for the Ce, Gd doped ZnO nanostructures [34–36]. In addition, the photocatalytic efficiency may also be due to the microstructural defects that arise in the nanostructures after doping/loading with other materials. Table 4 presents the main works on Mo-doped ZnO materials present in the literature regarding photocatalytic activity for the degradation of different dyes. These are compared to the materials (ZnO:MoO$_3$) presented in this work. According to the data listed in Table 4 [14,27,36,37], ZnO:MoO$_3$ (with 3.26% molar ratio Mo/Zn into ZnO matrix) showed a good value of the reaction rate constants ($k = 0.1072$ h^{-1} or 0.0018 min^{-1}) related to other studies for similar materials. However, the advantages of these materials are degradation of dyes in visible light of low intensity, the use of a reasonable amount of catalyst (0.5 g/L), and testing the material in soft conditions without acidification of solutions, in the absence of H$_2$O$_2$, and degradation of dyes at room temperature, usually used to boost the photochemical reactions.

Table 4. Comparison of photocatalytic activities of ZnO:Mo materials.

Photocatalyst Type	Type and Concentration of Dyes	Amount Photocatalyst (g/mL)	Light Source	Reaction Rate Constant k (min^{-1})	Ref.
ZnO@h-MoO$_3$	Methylene Blue	0.5	Vis	-	[14]
ZnO:Mo (2%)	Orange II	0.05/80	UV	0.0032	[27]
ZnO:Mo (0.6%)	Direct Yellow 27 Acid Blue 129	1/1000 1/1000	Vis (500 W)	0.0007 0.00085	[36]
MoO$_3$/ZnO	Methylene Blue	-	UV	0.00138	[37]
ZnO:MoO$_3$	Rhodamine B	0.005/10	Vis (100 W)	0.0018	This work

4. Conclusions

Pure ZnO, MoO$_3$ and MoO$_3$-incorporated ZnO nanostructures were prepared by electrospinning and calcination at 500 °C for 2 h. The XRD diffractograms confirmed the hexagonal wurtzite-type structure for pure ZnO, the orthorhombic crystalline structure obtained for MoO$_3$, and a combination of these structures for the ZnO:MoO$_3$ (S5) composite nanostructure. The optical properties of the prepared MoO$_3$-incorporated ZnO samples were studied using UV–vis absorption spectroscopy, and steady-state/time-resolved fluorescence spectroscopy. The absorption and emission bands are blue-shifted as MoO$_3$ concentrations increase. The fluorescence lifetime decay analysis exhibits a bi-exponential equation, indicating the existence of the energy transfer processes, and the values of the average emission lifetime decrease with increasing MoO$_3$ concentration. Moreover, the excited-state dynamics of S3 and S5 were characterized by nanosecond transient absorption spectroscopy. The photocatalytic activity of the pure/MoO$_3$-incorporated ZnO nanostructures obtained was tested for the use in photocatalytic degradation of the aqueous rhodamine B dye solution under visible light irradiation. The values of the reaction rate constant obtained for $ZnO:MoO_3$ sample (S5) was higher as compared to that of pure ZnO (S1) and depends on the MoO_3 concentration (upon the incorporation of 3.26% molar ratio Mo/Zn into ZnO matrix, the value of reaction rate constant doubled). The present contribution is the first report on synthesis of these materials by electrospinning followed by calcination as well as study of photocatalytic activity under visible light for rhodamine B dye degradation.

Author Contributions: Author Contributions: P.P., and O.I.; formal analysis, P.P. and O.I.; investigation, A.A., P.P., M.H. and N.O.; resources, O.I.; data curation, P.P. and O.I.; writing—original draft preparation, P.P. and O.I.; writing—review and editing, P.P. and O.I. All authors have read and agreed to the published version of the manuscript.

Funding: This research received no external funding.

Conflicts of Interest: The authors declare no conflict of interest.

References

1. Pascariu, P.; Homocianu, M. ZnO-Based ceramic nanofibers: Preparation, properties and applications. *Ceram. Int.* **2019**, *45*, 11158–11173. [CrossRef]

2. Türkyılmaz, Ş.Ş.; Güy, N.; Özacar, M. Photocatalytic efficiencies of Ni, Mn, Fe and Ag doped ZnO nanostructures synthesized by hydrothermal method: The synergistic/antagonistic effect between ZnO and metals. *J. Photochem. Photobiol. A Chem.* **2017**, *341*, 39–50. [CrossRef]

3. Pascariu, P.; Airinei, A.; Olaru, N.; Olaru, L.; Nica, V. Photocatalytic degradation of Rhodamine B dye using ZnO–SnO_2 electrospun ceramic nanofibers. *Ceram. Int.* **2016**, *42*, 6775–6781. [CrossRef]

4. Pascariu, P.; Tudose, I.V.; Suchea, M.P.; Koudoumas, E.; Fifere, N.; Airinei, A. Preparation and characterization of Ni, Co doped ZnO nanoparticles for photocatalytic applications. *Appl. Surf. Sci.* **2018**, *448*, 481–488. [CrossRef]

5. Chithambararaj, A.; Sanjini, N.S.; Velmathi, S.; Bose, A.C. Preparation of h-MoO3 and α-MoO3 nanocrystals: Comparative study on photocatalytic degradation of methylene blue under visible light irradiation. *Phys. Chem. Chem. Phys.* **2013**, *15*, 14761–14769. [CrossRef]

6. Wei, W.; Zhang, Z.; You, G.; Shan, Y.; Xu, Z. Preparation of recyclable MoO3 nanosheets for visible-Light driven photocatalytic reduction of Cr(vi). *RSC Adv.* **2019**, *9*, 28768–28774. [CrossRef]

7. Peña-Bahamonde, J.; Wu, C.; Fanourakis, S.K.; Louie, S.M.; Bao, J.; Rodrigues, D.F. Oxidation state of Mo affects dissolution and visible-Light photocatalytic activity of MoO3 nanostructures. *J. Catal.* **2020**, *381*, 508–519. [CrossRef]

8. Bai, S.; Chen, S.; Chen, L.; Zhang, K.; Luo, R.; Li, D.; Liu, C.C. Ultrasonic synthesis of MoO3 nanorods and their gas sensing properties. *Sens. Actuators B Chem.* **2012**, *174*, 51–58. [CrossRef]

9. Al-Otaibi, A.L.; Ghrib, T.; Alqahtani, M.; Alharbi, M.A.; Hamdi, R.; Massoudi, I. Structural, optical and photocatalytic studies of Zn doped MoO3 nanobelts. *Chem. Phys.* **2019**, *525*, 110410. [CrossRef]

10. Ong, C.B.; Ng, L.Y.; Mohammad, A.W. A review of ZnO nanoparticles as solar photocatalysts: Synthesis, mechanisms and applications. *Renew. Sustain. Energy Rev.* **2018**, *81*, 536–551. [CrossRef]

11. Kim, B.-Y.; Yoon, J.-W.; Lee, C.-S.; Park, J.-S.; Lee, J.-H. Trimethylamine Sensing Characteristics of Molybdenum doped ZnO Hollow Nanofibers Prepared by Electrospinning. *J. Sens. Sci. Technol.* **2015**, *24*, 419–422. [CrossRef]

12. Qamar, M.T.; Aslam, M.; Rehan, Z.; Soomro, M.T.; Ahmad, I.; Ishaq, M.; Ismail, I.M.; Fornasiero, P.; Hameed, A. MoO3 altered ZnO: A suitable choice for the photocatalytic removal of chloro-Acetic acids in natural sunlight exposure. *Chem. Eng. J.* **2017**, *330*, 322–336. [CrossRef]

13. Li, G.R.; Wang, Z.L.; Zheng, F.L.; Ou, Y.N.; Tong, Y.X. ZnO@MoO3 core/shell nanocables: Facile electrochemical synthesis and enhanced supercapacitor performances. *J. Mater. Chem.* **2011**, *21*, 4217–4221. [CrossRef]

14. Saminathanab, D.; Deogratiasa, T.; Thirugnanasambandana, S.; Vengidusamyc, N.; Arumainathana, S.; Dhanavel, S.; Sivaranjani, T.; Narayanan, V.; Stephen, A. Facile synthesis of 1D/1D ZnO@h-MoO3 for enhanced visible light driven photo degradation of industrial textile effluent. *Mater. Lett.* **2020**, *262*, 127049. [CrossRef]

15. Cojocaru, C.; Dorneanu, P.P.; Airinei, A.; Olaru, N.; Samoila, P.; Rotaru, A. Design and evaluation of electrospun polysulfone fibers and polysulfone/NiFe2O4 nanostructured composite as sorbents for oil spill cleanup. *J. Taiwan Inst. Chem. Eng.* **2017**, *70*, 267–281. [CrossRef]

16. Dorneanu, P.P.; Cojocaru, C.; Samoila, P.; Olaru, N.; Airinei, A.; Rotaru, A. Novel fibrous composites based on electrospun PSF and PVDF ultrathin fibers reinforced with inorganic nanoparticles: Evaluation as oil spill sorbents. *Polym. Adv. Technol.* **2018**, *29*, 1435–1446. [CrossRef]

17. Faraz, M.; Naqvi, F.K.; Shakir, M.; Khare, N. Synthesis of samarium-Doped zinc oxide nanoparticles with improved photocatalytic performance and recyclability under visible light irradiation. *New J. Chem.* **2018**, *42*, 2295–2305. [CrossRef]

18. Jia, T.; Wang, W.; Long, F.; Zhengyi, F.; Wang, H.; Zhang, Q. Fabrication, characterization and photocatalytic activity of La-Doped ZnO nanowires. *J. Alloys Compd.* **2009**, *484*, 410–415. [CrossRef]

19. Shaban, M.; Zayed, M.; Hamdy, H. Nanostructured ZnO thin films for self-Cleaning applications. *RSC Adv.* **2017**, *7*, 617–631. [CrossRef]

20. Srikant, V.; Clarke, D.R. On the optical band gap of zinc oxide. *J. Appl. Phys.* **1998**, *83*, 5447–5451. [CrossRef]

21. Tu, Y.F.; Huang, S.Y.; Sang, J.P.; Zou, X.W. Synthesis and photocatalytic properties of Sn-Doped TiO2 nanotube arrays. *J. Alloys Compd.* **2009**, *482*, 382–387. [CrossRef]

22. Neena, D.; Kondamareddy, K.K.; Bin, H.; Lu, D.; Kumar, P.; Dwivedi, R.K.; Pelenovich, V.O.; Zhao, X.-Z.; Gao, W.; Fu, D. Enhanced visible light photodegradation activity of RhB/MB from aqueous solution using nanosized novel Fe-Cd co-Modified ZnO. *Sci. Rep.* **2018**, *8*, 10691.

23. Xiao, F.; Chen, R.; Shen, Y.Q.; Dong, Z.L.; Wang, H.H.; Zhang, Q.Y.; Sun, H. Efficient Energy Transfer and Enhanced Infrared Emission in Er-Doped ZnO-SiO2 Composites. *J. Phys. Chem. C* **2012**, *116*, 13458–13462. [CrossRef]

24. Pandey, P.; Kurchania, R.; Haque, F.Z.; Khan, F.Z. Structural, diffused reflectance and photoluminescence study of cerium doped ZnO nanoparticles synthesized through simple sol–Gel method. *Optik* **2015**, *126*, 3310–3315. [CrossRef]

25. Fang, Y.; Lang, J.; Wang, J.; Han, Q.; Zhang, Z.; Zhang, Q.; Yang, J.; Zhong, B. Rare-Earth doping engineering in nanostructured ZnO: A new type of eco-Friendly photocatalyst with enhanced photocatalytic characteristics. *Appl. Phys. A* **2018**, *124*, 1–12. [CrossRef]

26. Theyvaraju, D.; Muthukumaran, S. Preparation, structural, photoluminescence and magnetic studies of Cu doped ZnO nanoparticles co-Doped with Ni by sol–Gel method. *Physica E* **2015**, *74*, 93–100. [CrossRef]

27. Mondal, P. Effect of Oxygen vacancy induced defect on the optical emission and excitonic lifetime of intrinsic ZnO. *Opt. Mater.* **2019**, *98*, 109476. [CrossRef]

28. Yu, C.; Yang, K.; Shu, Q.; Yu, J.C.-M.; Cao, F.; Li, X.; Zhou, X. Preparation, characterization and photocatalytic performance of Mo-Doped ZnO photocatalysts. *Sci. China Ser. B Chem.* **2012**, *55*, 1802–1810. [CrossRef]

29. McCamy, C.S. Correlated color temperature as an explicit function of chromaticity coordinates. *Color Res. Appl.* **1992**, *17*, 142–144. [CrossRef]

30. Hernández-Andrés, J.; Lee, R.L.; Romero, J. Calculating correlated color temperatures across the entire gamut of daylight and skylight chromaticities. *Appl. Opt.* **1999**, *38*, 5703–5709. [CrossRef] [PubMed]

31. Phuruangrat, A.; Yayapao, O.; Thongtem, T.; Thongtem, S. Preparation, characterization and photocatalytic properties of Ho doped ZnO nanostructures synthesized by sonochemical method. *Superlattices Microstruct.* **2014**, *67*, 118–126. [CrossRef]

32. Franco, A., Jr.; Pessoni, H.V.S. Optical band-Gap and dielectric behavior in Ho-Doped ZnO nanoparticles. *Mater. Lett.* **2016**, *180*, 305–308. [CrossRef]

33. Hirotaka, N.; Koichi, K.; Masashi, T. Fabrication and photocatalytic activity of TiO2/MoO3 particulate films. *J. Oleo Sci.* **2009**, *58*, 203–211.

34. Lang, J.; Wang, J.; Zhang, Q.; Li, X.; Han, Q.; Wei, M.; Sui, Y.; Wang, D.; Yang, J. Chemical precipitation synthesis and significant enhancement in photocatalytic activity of Ce-Doped ZnO nanoparticles. *Ceram. Int.* **2016**, *42*, 14175–14181. [CrossRef]

35. George, A.; Sharma, S.K.; Chawla, S.; Malik, M.M.; Qureshi, M. Detailed of X-Ray diffraction and photoluminescence studies of Ce doped ZnO nanocrystals. *J. Alloys Compd.* **2011**, *509*, 5942–5946. [CrossRef]

36. Franco, A., Jr.; Pessoni, H.V.S. Effect of Gd doping on the structural, optical band-Gap, dielectric and magnetic properties of ZnO nanoparticles. *Phys. B Condens. Matter* **2017**, *506*, 145–151. [CrossRef]

37. Umar, K.; Aris, A.; Parveen, T.; Jaafar, J.; Majid, Z.A.; Reddy, A.V.B.; Talib, J. Synthesis, characterization of Mo and Mn doped ZnO and their photocatalytic activity for the decolorization of two different chromophoric dyes. *Appl. Catal. A Gen.* **2015**, *505*, 507–514. [CrossRef]

Article

3D Printed Fully Recycled TiO$_2$-Polystyrene Nanocomposite Photocatalysts for Use against Drug Residues

Maria Sevastaki [1,2], Mirela Petruta Suchea [3,4,*] and George Kenanakis [1,*]

[1] Institute of Electronic Structure and Laser, Foundation for Research & Technology-Hellas, N. Plastira 100, 70013 Heraklion, Crete, Greece; msevastaki@iesl.forth.gr
[2] Department of Chemistry, University of Crete, 71003 Heraklion, Crete, Greece
[3] National Institute for Research and Development in Microtechnologies (IMT-Bucharest), 1 26 A, Erou Iancu Nicolae Street, P.O. Box 38-160, 023573 Bucharest, Romania
[4] Center of Materials Technology and Photonics, Hellenic Mediterranean University, 71004 Heraklion, Crete, Greece
* Correspondence: mira.suchea@imt.ro (M.P.S.); gkenanak@iesl.forth.gr (G.K.)

Received: 12 October 2020; Accepted: 26 October 2020; Published: 28 October 2020

Abstract: In the present work, the use of nanocomposite polymeric filaments based on 100% recycled solid polystyrene everyday products, enriched with TiO$_2$ nanoparticles with mass concentrations up to 40% *w/w*, and the production of 3D photocatalytic structures using a typical fused deposition modeling (FDM)-type 3D printer are reported. We provide evidence that the fabricated 3D structures offer promising photocatalytic properties, indicating that the proposed technique is indeed a novel low-cost alternative route for fabricating large-scale photocatalysts, suitable for practical real-life applications.

Keywords: metal oxide nano-structures; titanium dioxide (TiO$_2$); photocatalysis; paracetamol; 3D printing; fused filament fabrication (FFF); fused deposition modeling (FDM)

---

1. Introduction

For many years, human activities have polluted the environment in many ways; several organic residuals originating from highly toxic pollutants such as pharmaceuticals can be found in water, comprising a critical health and environmental issue [1–3]. In many cases, unused, expired and residual pharmaceuticals are discharged into the sewerage system, burdening the aquifer. Moreover, these compounds can also be discharged into the environment through the metabolism of human bodies [4–8]. As a result, pharmaceuticals have been found in sewage, surface and ground water in many countries [6,8,9].

The most common methods used to overcome this problem include the return of medicines to pharmacies and not throwing expired medicinal products in the sewer, as well as biological degradation, chlorination or ozonation, but these are not efficient enough to remove these compounds from the treated water [10–12]. These drugs residues must be eliminated using an oxidation method, and advanced heterogeneous photocatalysis seems to be one of the most promising approaches, since it implies the use of an inert catalyst, non-hazardous oxidants and ultraviolet (UV) and/or visible light input [13–25].

Heterogeneous photocatalysis using TiO$_2$ is a method generating free radical •OH using atmospheric air instead of O$_3$ or H$_2$O$_2$, significantly reducing processing costs. The process takes place at ambient conditions and leads to the complete decomposition of both liquid and gaseous pollutants [26,27]. The greatest advantage of this method is that an environmentally friendly catalyst which is widely available, inexpensive, non-toxic, and photo-stable with respect to other photocatalysts is readily available and readily regenerable for the purpose of reuse, maintaining equally high performance for a large number of catalytic cycles [13–17,26]. Several studies have shown that titanium dioxide (TiO_2) is the most potent semiconductor for the oxidative destruction of organic compounds. TiO_2 has, in addition to its large photocatalytic activity, greater resistance to corrosion and photo-corrosion, resulting in the possibility of recycling [28].

The disadvantage of photocatalysis when the semiconductor is used in the form of a powder is the need to remove it after the end of the treatment [29,30]. For this reason, international efforts are focused on photocatalytic systems, where the catalyst is used in the form of a film on inert substrates to eliminate the stage of powder removal [31–35]. Over and above that, photocatalytic efficiency increases with effective surface area, and consequently a nanostructured photocatalyst is beneficial. However, solid catalysts' samples, such as thin films or nanostructured ones, in most cases cannot exceed an overall size of a few centimeters, due to the limitation of the fabrication techniques, limiting their potential use in real-life applications.

In the last few years, 3D printing technology become of great interest in several fields of research, such as medicine, chemistry and materials science, as an effective, fancy, quick and low-cost route for the production of 3D large-size samples [36–40]. The most common technique is fused deposition modeling (FDM) in which polymers are the usual materials used as filaments. It should be noted that although there are several reports on 3D structures for novel environmental applications [41–43], there are only a few in which custom-made filaments (with nanoparticles of inorganic materials into a polymeric matrix) are used in combination with FDM technology, i.e., in [44,45], and none for drug-residuals' removal by photocatalysis.

This work discusses an investigation of the photocatalytic degradation of paracetamol (also known as acetaminophen, APAP), a medicine available in a huge number of countries worldwide, used to treat pain and fever, using 3D-printed photocatalysts enriched with 20% *w/w* nanostructured TiO_2. It is worth mentioning that the polymeric filaments used to produce these 3D-printed photocatalysts are based on 100% recycled solid polystyrene (PS) everyday products, such as containers, lids, CD cases etc., following an eco-friendly environmental approach. APAP was chosen for this study due to its high occurrence as a pharmaceutical pollutant in environment. As shown in many studies, it was found worldwide in almost all kinds of water source as well in the soil [46]. APAP is reported to be one of the most frequently detected pharmaceuticals in sewage treatment plant effluents [47]. The increasing concentrations of APAP together with other emerging contaminants result in the occurrence of toxic phenomena in non-target species present in receiving aquatic environments. An excellent review regarding the toxic effects of environmental APAP is presented in Ref. [48].

The present experimental results provided strong evidence that the proposed fully recycled 3D printed photocatalysts are good candidates against the degradation of APAP drug residues, reaching an efficiency of almost 75% of a 100 ppm APAP aqueous solution under UV irradiation for 20 min, and ~60% after three cycles of reuse in 200 ppm APAP aqueous solutions, respectively.

2. Experimental Details

2.1. Synthesis of the Metal Oxide Polymeric Nanocomposites

First, several everyday PS products, such as containers, lids, CD cases etc. were ground using an IKA A11 Basic Analytical Mill (IKA-Werke GmbH and Co. KG, Staufen, Germany) equipped with a high-grade stainless-steel beater, coated with chromium carbide. Recycled grinded PS powder/beads of ~0.2 mm diameter were dissolved in toluene (Merck KGaA, Darmstadt, Germany) (in a sealed bottle, under continuous stirring for 2 h) to create a 20% *w/w* solution. The resultant solution was stirred for 24 h at room temperature using a magnetic stirrer to yield a homogeneous, milky solution.

Subsequently, 2 g and 4 g of commercially available TiO_2 nanoparticles (TiO_2 P25 with a mean particle size of ~25 nm, obtained from Evonic Industries AG, Essen, Germany) were introduced in

10 mL of the PS/toluene solution mentioned above under stirring at 40 °C for 30 min, in order to obtain the TiO_2 homogeneous suspensions with 20% *w/w* and 40% *w/w* concentration in PS, respectively.

Each of the TiO_2 homogeneous suspensions was transferred to 200 mL of ethanol (95% purity; Merck KGaA, Darmstadt, Germany), to form a dense precipitate, which consisted of the PS and the suspended metal oxide nanoparticles. After formation, the precipitate was collected and dried at 60 °C for 24 h using a typical laboratory oven (Memmert UNP 500 Memmert GmbH + Co., Schwabach, Germany). Using the procedure above, 20 g of 20% *w/w* TiO_2/PS, and 20 g 40% *w/w* TiO_2/PS solid nanocomposite solutions were produced, respectively.

2.2. Filament Production

The produced TiO_2/PS solid nanocomposite solutions were cut in ~2–3 mm^2 pieces, and forwarded to a "Noztek Pro" (Noztek, Shoreham, West Sussex, UK) high temperature extruder, and processed at 240 °C, in order to transform them to a cylindrical filament with a diameter of 1.75 ± 0.15 mm, suitable for 3D-printing. All extrusion parameters, such as extrusion velocity and temperature, were optimized towards the production of a uniform continuous cylindrical cord, with an overall length of ~5 m.

2.3. Production of 3D-Printed Photocatalytic Structures

Flat, rectangular-shaped (10 mm × 10 mm × 1 mm) 3D structures were designed using "Tinkercad", a free online 3D design and 3D printing software from Autodesk Inc (Mill Valley, CA, USA). A dual-extrusion FDM-type 3D printer (Makerbot Replicator 2X; MakerBot Industries, Brooklyn, NY, USA) was used for the direct fabrication of TiO_2/PS nanocomposite photocatalytic samples, using the PS/TiO_2 nanocomposite filaments described above. The FDM process of building a solid object involves heating of the fed filament and pushing it out layer-by-layer through a heated (240 °C) nozzle (0.4 mm inner diameter) onto a heated surface (80 °C), via a computer controlled three-axis positioning system (with a spatial resolution of approximately 100 μm in the z-axis and 11 μm in x and y).

2.4. Characterization and Photocatalytic Experiments

X-ray diffraction (XRD) measurements were performed in order to determine the crystalline structure of the 3D-printed samples, using a Rigaku RINT 2000 (Rigaku, Tokyo, Japan) diffractometer with Cu Kα (λ = 1.5406Å) X-rays for 2θ = 20.00–60.00° for TiO_2/PS nanocomposite-based samples and a step time 60°/sec.

Furthermore, Raman spectroscopy measurements were performed at room temperature using a Horiba LabRAM HR Evolution (HORIBA FRANCE SAS, Longjumeau, France) confocal micro-spectrometer, in backscattering geometry (180°), equipped with an air-cooled solid-state laser operating at 532 nm with 100 mW output power. The laser beam was focused on the samples using a 10× Olympus (OLYMPUS corporation, Tokyo, Japan) microscope objective (numerical aperture of 0.25), providing ~14 mW power on each sample. Raman spectra over the 100–700 cm^{-1} wavenumber range (with an exposure time of 5 s and 3 accumulations) were collected by a Peltier cooled CCD (1024 × 256 pixels) detector (HORIBA FRANCE SAS, Longjumeau, France) at −60 °C, with a resolution better than 1 cm^{-1}, achieved thanks to an 1800 grooves/mm grating and an 800 mm focal length. Test measurements carried out using different optical configuration, exposure time, beam power and accumulations in order to obtain sufficiently informative spectra using a confocal hole of 100 μm, but ensuring to avoid alteration of the sample, while the high spatial resolution allowed us to carefully verify the sample homogeneity. The wavelength scale was calibrated using a Silicon standard (520.7 cm^{-1}) (Silchem Handelsgesellschaft mbH, Freiberg, Germany) and the acquired spectra were compared with scientific published data and reference databases, such as Horiba LabSpec 6 (HORIBA FRANCE SAS, Longjumeau, France).

The photocatalytic activity of the 3D-printed samples was studied by means of the reduction of APAP in aqueous solution, which is a well-known pharmaceutical product that has been used as a model organic to probe the photocatalytic performance of photocatalysts [4,5,7,8]. The investigated samples were placed in a vertical custom-made quartz cell, and the whole setup (cell + solution + sample) was illuminated up to 60 min using an HPK 125 W Philips UV lamp centered at 365 nm (msscientific Chromatographie-Handel GmbH, Berlin, Germany) with a light intensity of ~6.0 mW/cm^2. The concentration of APAP (degradation) was monitored by UV-Vis spectroscopy in absorption mode (absorption at λ_{max}, 665 nm), using a K-MAC SV2100 (K-MAC, Daejeon, Korea) spectrophotometer over the wavelength range of 220–800 nm. In such way, UV-Vis absorption data were collected at 0, 10, 20, 30 and 40 min, while the quantification of the APAP removal (and hence the remaining APAP concentration) was estimated by calculation of the area below the main APAP peak in the range of 220–320 nm. Additional blank experiments (photolysis) without a catalyst were also performed as well as APAP adsorption experiments in the dark.

3. Results and Discussion

In order to verify the nominal TiO_2 loading in PS filaments and 3D-printed nanocomposites, a type of thermogravimetric method was used. A small piece of each sample on a quartz substrate, was weighted and was heated at ~900 °C in order to burn all organics and polymeric residuals, then weighed again. Since TiO_2 is not affected at all at such temperatures, the remaining mass was the TiO_2 loading. This way we checked that the nominal TiO_2 % *w/w* loadings were indeed 20% and 40% *w/w* ± 0.5–1.0%.

Figure 1 depicts a typical optical microscopy photograph of a 3D printed sample (40% *w/w* TiO_2/PS), as fabricated following the FDM process mentioned above.

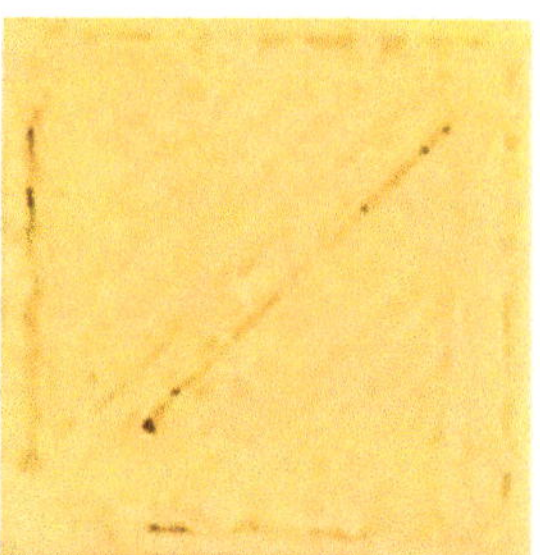

Figure 1. Typical photograph of a 3D-printed nanocomposite photocatalytic sample with 40% *w/w* TiO_2 in polystyrene (PS).

As one can notice from Figure 1, rough structures were printed instead of smooth ones, while printing directions were also observed. In our case, the nanoparticle loading (40% *w/w*, shown in Figure 1) in the custom-made filaments, most likely led to low-resolution/low-quality 3D printing. It hence became clear that further investigation was needed in order to improve the printing quality.

Figure 2 presents typical XRD patterns for PS/TiO$_2$ 3D printed structures. Well-distinguished diffraction peaks are observed. These correspond to both anatase and rutile phase, in good agreement with the JCPDS card (No. 84–1286) and JCPDS card (No. 88–1175), a crystal structure of anatase and rutile, respectively [49,50] normal for P25 Degussa TiO$_2$ that is a mix of the two phases.

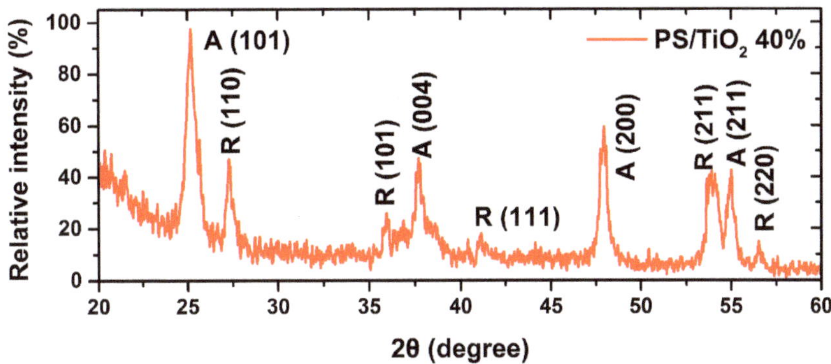

Figure 2. Typical X-ray diffraction (XRD) patterns for PS/TiO$_2$ 3D-printed nanocomposite structures.

Figure 3 shows a typical Raman spectrum of the PS/TiO$_2$ 3D printed structures, which exhibit characteristic TiO$_2$ phonon frequencies, such as: 143 cm^{-1} (E$_g$), 396 cm^{-1} (B$_{1g}$), 516 cm^{-1} (A$_{1g}$) for anatase, and 245 cm^{-1} (two-phonon scattering) and 610 cm^{-1} (A$_{1g}$) for rutile, matching ($\pm$ 3 cm^{-1}) with literature [51–53].

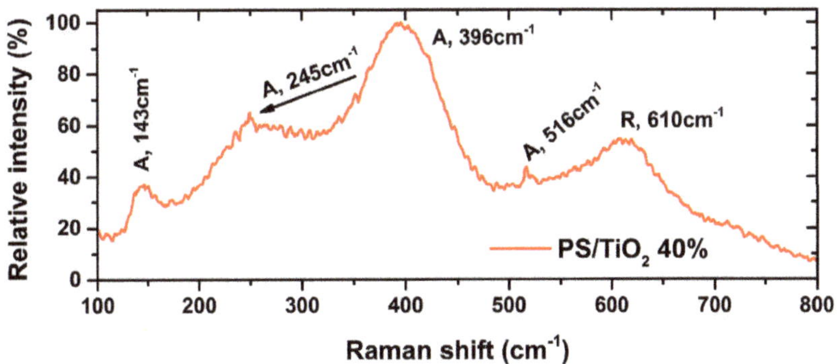

Figure 3. Typical Raman spectra for PS/TiO$_2$ 3D-printed nanocomposite structures.

The photocatalytic activity of the 3D-printed nanocomposites under UV-A light was evaluated by assessing the reduction of APAP in aqueous solution. The photolytic removal (photolysis) of the pharmaceutical product (in the absence of any photocatalyst) was negligible, underlining the indispensability of the catalysts. Furthermore, to eliminate the possibility of APAP removal by adsorption on the catalysts, the samples were placed at the bottom of the reactor under dark conditions and in contact with the APAP for 30 min, during which time equilibrium of adsorption-desorption was reached. In all cases, removal was insignificant (less than 3%), pointing to the fact that the reduction of the APAP should be attributed to a pure photocatalytic procedure.

The decrease of the concentration of APAP (20 ppm) using both 20% *w/w* and 40% *w/w* 3D-printed TiO$_2$/PS nanocomposite samples under UV-A light irradiation is presented in Figure 4. For comparison reasons, the photolysis curve (no catalyst present) is also displayed. According to the photolysis (black curve in Figure 4), the concentration of APAP remained almost constant during ~40 min irradiation, indicating that the photolysis of APAP was almost negligible.

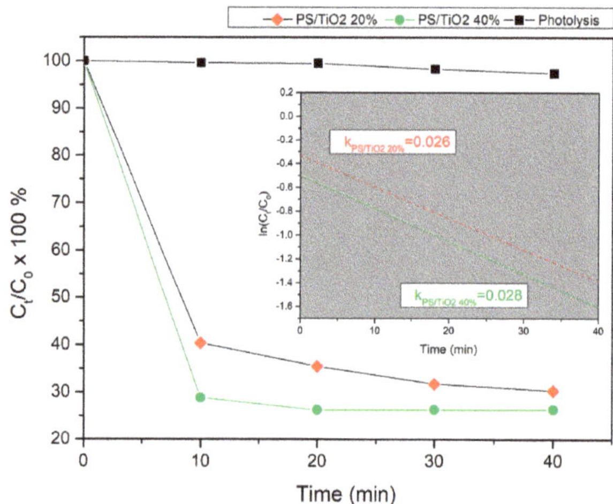

Figure 4. Percentage (%) acetaminophen (APAP) degradation using 20% *w/w* and 40% *w/w* [red solid rhombuses and green solid circles] TiO$_2$-based 3D-printed nanocomposites under ultraviolet (UV-A) irradiation, vs. irradiation time, respectively. For comparison reasons, the photolysis curve (black solid squares) is also presented. In the inset one can see the apparent rate constants (k) of APAP degradation using 20% *w/w* and 40% *w/w* 3D printed TiO$_2$/PS nanocomposite photocatalysts.

As shown in Figure 4, the 40% *w/w* 3D-printed TiO$_2$/PS nanocomposite photocatalysts were highly effective regarding the reduction of APAP compared to the 20% *w/w* 3D printed TiO$_2$/PS nanocomposite ones, due to the highly oxidative radicals generated on the TiO$_2$ at the surfaces under UV-A irradiation [51].

As already stated, (and shown in the inset of Figure 4), the photodegradation of APAP using the 3D-printed TiO$_2$/PS nanocomposite samples followed a first-order kinetics. The calculated apparent rate constants were 0.026 min^{-1} and 0.028 min^{-1} for 20% *w/w* and 40% *w/w* 3D-printed TiO$_2$/PS nanocomposite samples, respectively. One can notice that the 40% *w/w* 3D-printed TiO$_2$/PS nanocomposite samples are more photocatalytically active than the 20% *w/w* ones, regarding the degradation of APAP, reaching an almost 30% reduction of APAP's concentration after 10 min of irradiation.

In principle, when a semiconductor is exposed to electromagnetic radiation of an appropriate wavelength, excitation occurs and electrons (e$_{CB-}$) are transferred from the valence band to the conduction band of the material, leaving behind positively charged holes (h$_{VB+}$). The photogenerated holes react with OH$^-$ or H$_2$O adsorbed on the surface of the catalyst, and hydroxyl radicals that are mainly responsible for the degradation of the target pollutant are created. It is therefore expected that a high recombination rate of photogenerated holes and electrons will be disadvantageous for the performance of the photocatalyst.

Nevertheless, an efficient electron and hole transfer between TiO$_2$ depends on the difference between the conduction and valence band potentials of the semiconductor, that should be suitably positioned [54,55]. Concentration of catalysts in the 40% *w/w* 3D-printed TiO$_2$/PS nanocomposite is double that in the 20% *w/w* 3D-printed TiO_2/PS nanocomposite samples thus allowing charge separation and increasing the efficiency of the photocatalytic reaction.

In addition, the apparent rate constant (k) has been calculated as the basic kinetic parameter for the comparison of photocatalytic activities, which was fitted by the equation $ln(C_t/C_0) = -kt$, where k is apparent rate constant, C_t is the concentration of APAP, and C_0 is the initial concentration of APAP. It should be noted that the adjusted R-square statistic varies from 0.91499 to 0.94334 indicating that the model used for the determination of the apparent rate constant *(k)* is adequate. The good linear fit of equation $ln(C_t/C_{,0}) = -kt$ shown in the inset of Figure 4, confirms that the photodegradation for all different concentrations of APAP using 3D printed TiO_2/PS nanocomposite photocatalysts at 20% and 40% *w/w*, follows first-order kinetics.

It is worth mentioning that the photocatalytic activity tests were carried out at least three times on the 3D-printed TiO_2/PS nanocomposite samples to examine their stability under UV illumination, demonstrating no changes in the photocatalytic activity after three runs. Moreover, at least three structures with the same TiO_2 load have been produced and tested, in order to check the reproducibility of the structure manufacturing.

Furthermore, the photocatalytic activity of 40% *w/w* 3D-printed TiO_2/PS nanocomposite was checked in APAP aqueous solutions with different concentrations (from 20 ppm to 200 ppm). As can be observed from Figure 5a, when the concentration of APAP increases, more irradiation time is needed for its degradation. Although this is expected, it is worth mentioning that for a 100 ppm APAP solution, 20 min of irradiation are enough to reduce it at ~75%, while for 200 ppm APAP solution, 40 min are enough in order to reduce it by the same amount.

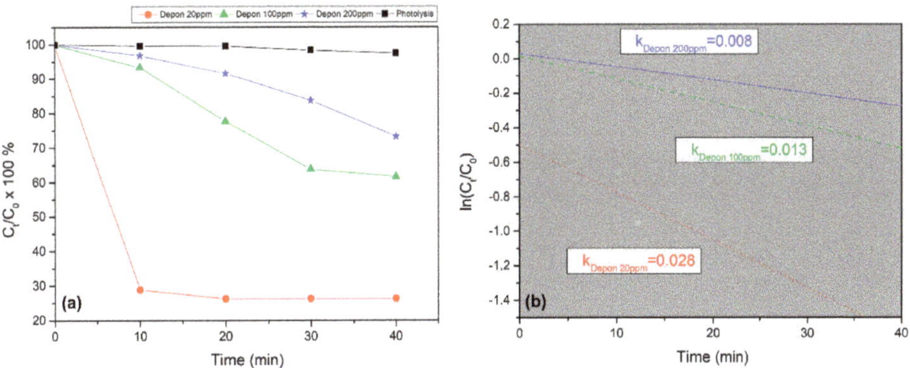

Figure 5. (**a**) % APAP degradation using 40% *w/w* 3D-printed TiO_2/PS nanocomposites under UV-A irradiation, vs. irradiation time. Three different concentrations of APAP are presented; 20 ppm, 100 ppm and 200 ppm (red solid circles, green solid triangles and blue solid stars, respectively). (**b**) The apparent rate constants (k) of APAP degradation (20 ppm, 100 ppm, and 200 ppm, respectively), using 40% *w/w* 3D printed TiO_2/PS nanocomposite photocatalysts.

Figure 5b confirms that the photodegradation of APAP using 3D printed TiO_2/PS nanocomposite samples follow the first-order kinetics. The constant k as was calculated (as described above) to be $0.008\ min^{-1}$, $0.013\ min^{-1}$ and $0.028\ min^{-1}$ for 200 ppm, 100 ppm and 20 ppm of APAP, respectively.

To verify the use of the 3D-printed TiO_2/PS nanocomposite photocatalysts for practical environmental applications, each sample was recovered, and their efficiency tested for at least three runs. Figure 6 depicts the re-use of one of the 40% *w/w* 3D-printed TiO_2/PS nanocomposite for three runs, against the degradation of 100 ppm APAP aqueous solution.

As one can see from Figure 6, the 3D-printed TiO_2/PS nanocomposite photocatalysts can be successfully used at least 3 times for the photodegradation of APAP, reaching an efficiency of ~60% at the end of the 3rd run.

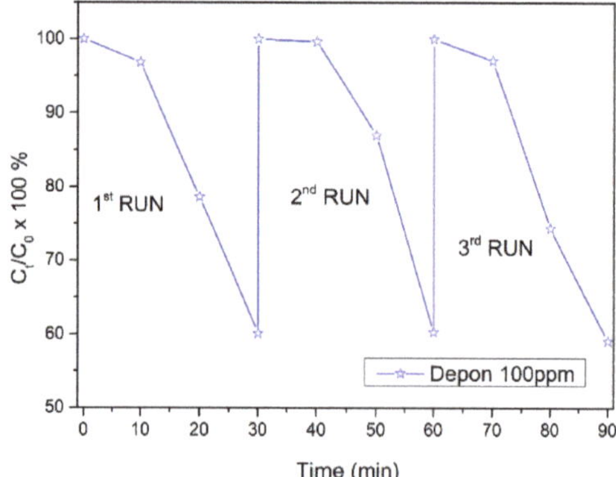

Figure 6. % APAP (100 ppm) degradation using a 40% *w/w* TiO$_2$/PS 3D-printed nanocomposite sample under UV-A irradiation, for 3 runs of 30 min irradiation each.

4. Summary and Conclusions

This work provides a novel experimental study concerning the successful use of TiO$_2$/PS nanocomposite polymeric filaments based on 100% recycled solid polystyrene everyday products, enriched with TiO$_2$ nanoparticles with mass concentrations up to 40%*w/w*, for the production of 3D photocatalytic structures/devices using a typical FDM-type 3D printer. The 3D-printed TiO$_2$/PS nanocomposites were successfully used as photocatalysts for the APAP degradation. It should be noted that this is the first report of 3D-printed photocatalytic devices made of fully recycled raw materials, and with a TiO$_2$ loading as high as high as 40% *w/w*.

The 3D-printed TiO$_2$/PS nanocomposite samples provide promising photocatalytic properties, reaching an efficiency of almost 60% after three cycles of reuse in 200 ppm of APAP aqueous solution under UV-A irradiation, offering a novel low-cost alternate way for fabricating large-scale photocatalysts, suitable for practical applications.

Author Contributions: G.K. developed the original concept. M.S. and G.K. performed the material synthesis and fabrication experiments. M.S., M.P.S. and G.K. carried out the structure characterization, and the photocatalysis experiments and data analysis and processing. G.K. supervised and coordinated the experimental work. All authors have read and agreed to the published version of the manuscript.

Funding: This work was supported by the National Priorities Research Program grant No. NPRP11S-1128-170042 from the Qatar National Research Fund (member of The Qatar Foundation), and co-financed by the European Union and Greek national funds through the Operational Program Competitiveness, Entrepreneurship and Innovation, under the call RESEARCH–CREATE–INNOVATE (acronym: POLYSHIELD; project code: T1EDK-02784).

Acknowledgments: M.P.S. contribution to this work was partially supported by the Ministry of Education and Research through Program 1—Development of the National R&D System, Subprogram 1.2—Institutional Performance—Projects for Excellence Financing in RDI, EXCEL-IMT, Contract no. 13 PFE/16.10.2018.

Conflicts of Interest: The authors declare that there are no conflicts of interest regarding the publication of this manuscript.

References

1. Kwon, C.H.; Shin, H.; Kim, J.H.; Choi, W.S.; Yoon, K.H. Degradation of methylene blue via photocatalysis of titanium dioxide. *Mater. Chem. Phys.* **2004**, *86*, 78–82. [CrossRef]
2. Arabatzis, M.; Antonaraki, S.; Stergiopoulos, T.; Hiskia, A.; Papaconstantinou, E.; Bernard, M.C.; Falaras, P. Preparation, characterization and photocatalytic activity of nanocrystalline thin film TiO$_2$ catalysts towards 3,5-dichlorophenol degradation. *J. Photochem. Photobiol. A Chem.* **2002**, *149*, 237–245. [CrossRef]
3. Alberici, R.M.; Jardim, W.F. Photocatalytic destruction of VOCs in the gas-phase using titanium dioxide. *Appl. Catal. B Environ.* **1997**, *14*, 55–68. [CrossRef]

4. Desale, A.; Kamble, S.P.; Deosarkar, M.P. Photocatalytic Degradation of Paracetamol Using Degussa TiO$_2$ Photocatalyst. *IJCPS* **2013**, *2*, 140–148.

5. Suryawanshia, M.A.; Maneb, V.B.; Kumbharc, G.B. Immersed Water Purifier: A Novel Approach towards Purification. *Int. J. Innov. Emerg. Res. Eng.* **2016**, *3*, 1–5.

6. Yang, L.; Yu, L.E.; Ray, M.B. Degradation of paracetamol in aqueous solutions by TiO$_2$ photocatalysis. *Water Res.* **2008**, *42*, 3480–3488. [CrossRef]

7. Shakir, M.; Faraz, M.; Sherwani, M.A.; Al-Resayes, S.I. Photocatalytic degradation of the Paracetamol drug using Lanthanum doped ZnO nanoparticles and their in-vitro cytotoxicity assay. *J. Lumin.* **2016**, *176*, 159–167. [CrossRef]

8. Thi, V.H.T.; Lee, B.K. Effective photocatalytic degradation of paracetamol using La-doped ZnO photocatalyst under visible light irradiation. *Mater. Res. Bull.* **2017**, *96*, 171–182. [CrossRef]

9. Cifci, D.I.; Tuncal, T.; Pala, A.; Uslu, O. Determination of optimum extinction wavelength for paracetamol removal through energy efficient thin film reactor. *J. Photochem. Photobiol. A Chem.* **2016**, *322–323*, 102–109.

10. Patil, S.S.; Shinde, V.M. Biodegradation studies of aniline and nitrobenzene in aniline plant wastewater by gas chromatography. *Environ. Sci. Technol.* **1988**, *22*, 1160–1165. [CrossRef]

11. More, A.T.; Vira, A.; Fogel, S. Biodegradation of trans-1,2-dichloroethylene by methane-utilizing bacteria in an aquifer simulator. *Environ. Sci. Technol.* **1989**, *23*, 403–406.

12. Slokar, Y.M.; Le Marechal, A.M. Methods of decoloration of textile wastewaters. *Dye. Pigment.* **1998**, *37*, 335–356.

13. Bahadur, N.; Jain, K.; Srivastava, A.K.; Govind, R.; Gakhar, D.; Haranath; Dulat, M.S. Effect of nominal doping of Ag and Ni on the crystalline structure and photo-catalytic properties of mesoporous titania. *Mater. Chem. Phys.* **2010**, *124*, 600–608.

14. Soutsas, K.; Karayannis, V.; Poulios, I.; Riga, A.; Ntampegliotis, K.; Spiliotis, X.; Papapolymerou, G. Decolorization and degradation of reactive azo dyes via heterogeneous photocatalytic processes. *Desalination* **2010**, *250*, 345–350.

15. Peng, Y.H.; Huang, G.F.; Huang, W.Q. Visible-light absorption and photocatalytic activity of Cr-doped TiO$_2$ nanocrystal films. *Adv. Powder Technol.* **2012**, *23*, 8–12.

16. Hashimoto, K.; Irie, H.; Fujishima, A. TiO$_2$ Photocatalysis: A Historical Overview and Future Prospects. *Jpn. J. Appl. Phys.* **2005**, *44*, 8269–8285.

17. Carp, O.; Huisman, C.L.; Reller, A. Photoinduced reactivity of titanium dioxide. *Prog. Solid State Chem.* **2004**, *32*, 33–177.

18. Kenanakis, G.; Giannakoudakis, Z.; Vernardou, D.; Savvakis, C.; Katsarakis, N. Photocatalytic degradation of stearic acid by ZnO thin films and nanostructures deposited by different chemical routes. *Catal. Today* **2010**, *151*, 34–38.

19. Chen, X.; Wu, Z.; Liu, D.; Gao, Z. Preparation of ZnO Photocatalyst for the Efficient and Rapid Photocatalytic Degradation of Azo Dyes. *Nanoscale Res. Lett.* **2017**, *12*, 1–10.

20. Frysali, M.A.; Papoutsakis, L.; Kenanakis, G.; Anastasiadis, S.H. Functional Surfaces with Photocatalytic Behavior and Reversible Wettability: ZnO Coating on Silicon Spikes. *J. Phys. Chem. C* **2015**, *119*, 25401–25407.

21. Yu, K.S.; Shi, J.Y.; Zhang, Z.L.; Liang, Y.M.; Liu, W. Synthesis, Characterization, and Photocatalysis of ZnO and Er-Doped ZnO. *J. Nanomater.* **2013**. [CrossRef]

22. Johar, M.A.; Afzal, R.A.; Alazba, A.A.; Manzoor, U. Photocatalysis and bandgap engineering using ZnO nanocomposites. *Adv. Mater. Sci. Eng.* **2015**. [CrossRef]

23. Kenanakis, G.; Katsarakis, N. Light-induced photocatalytic degradation of stearic acid byc-axisoriented ZnO nanowires. *Appl. Catal. A* **2010**, *378*, 227–233. [CrossRef]

24. Kenanakis, G.; Vernardou, D.; Katsarakis, N. Light-induced self-cleaning properties of ZnO nanowires grown at low temperatures. *Appl. Catal. A* **2012**, *411*, 7–14. [CrossRef]

25. Aguilar, C.A.; Montalvo, C.; Ceron, J.G.; Moctezuma, E. Photocatalytic Degradation of Acetaminophen. *Int. J. Environ. Res.* **2011**, *5*, 1071–1078.

26. Konstantinou, I.; Albanis, T. Photocatalytic Transformation of Pesticides in Aqueous Titanium Dioxide Suspensions Using Artificial and Solar Light: Intermediates and Degradation Pathways. *Appl. Catal. B* **2003**, *42*, 319–335. [CrossRef]

27. Tanaka, K.; Padermpole, K.; Hisanaga, T. Photocatalytic Degradation of Commercial Azo Dyes. *Water Res.* **2000**, *34*, 327–333. [CrossRef]

28. Mills, A.; Lee, S. A web-based overview of semiconductor photochemistry-based current commercial applications. *J. Photochem. Photobiol A Chem.* **2002**, *152*, 233–247. [CrossRef]

29. Li, Y.; Chen, J.; Liu, J.; Ma, M.; Chen, W.; Li, L. Activated carbon supported TiO_2-photocatalysis doped with Fe ions for continuous treatment of dye wastewater in a dynamic reactor. *J. Environ. Sci.* **2010**, *22*, 1290–1296. [CrossRef]

30. Van Grieken, R.; Marugán, J.; Sordo, C.; Martínez, P.; Pablos, C. Photocatalytic inactivation of bacteria in water using suspended and immobilized silver-TiO_2. *Appl. Catal. B Environ.* **2009**, *93*, 112–118. [CrossRef]

31. Banerjee, A.N.; Ghosh, C.K.; Chattopadhyay, K.K.; Minoura, H.; Sarkar, A.K.; Akiba, A.; Kamiya, A.; Endo, T. Low-temperature deposition of ZnO thin films on PET and glass substrates by DC-sputtering technique. *Thin Solid Films* **2006**, *496*, 112–116. [CrossRef]

32. Zhao, J.; Chen, C.; Ma, W. Photocatalytic Degradation of Organic Pollutants Under Visible Light Irradiation. *Top. Catal.* **2005**, *35*, 269–278. [CrossRef]

33. Mills, A.; Elliott, N.; Hill, G.; Fallis, D.; Durrant, J.R.; Willis, R.L. Preparation and characterisation of novel thick sol–gel titania film photocatalysts. *Photochem. Photobiol. Sci.* **2003**, *2*, 591–596. [CrossRef]

34. Henderson, M.A. A surface science perspective on TiO photocatalysis. *Surf. Sci. Rep.* **2011**, *66*, 185–297. [CrossRef]

35. Lopez, L.; Daoud, W.A.; Dutta, D. Preparation of large scale photocatalytic TiO_2 films by the sol-gel process. *Surf. Coat. Technol.* **2010**, *2*, 251–257. [CrossRef]

36. Beltrame, E.D.V.; Tyrwhitt-Drake, J.; Roy, I.; Shalaby, R.; Suckale, J.; Pomeranz Krummel, D. 3D Printing of Biomolecular Models for Research and Pedagogy. *J. Vis. Exp.* **2017**, *121*, 55427.

37. Ventola, C.L. Medical Applications for 3D Printing: Current and Projected Uses. *Pharm. Ther.* **2014**, *39*, 704–711.

38. Lee, J.Y.; An, J.; Chua, C.K. Fundamentals and applications of 3D printing for novel materials. *Appl. Mater. Today* **2017**, *7*, 120–133. [CrossRef]

39. Ambrosi, A.; Pumera, M. 3D-printing technologies for electrochemical applications. *Chem. Soc. Rev.* **2016**, *45*, 2740–2755. [CrossRef]

40. Kenanakis, G.; Vasilopoulos, K.C.; Viskadourakis, Z.; Barkoula, N.M.; Anastasiadis, S.H.; Kafesaki, M.; Economou, E.N.; Soukoulis, C.M. Electromagnetic shielding effectiveness and mechanical properties of graphite-based polymeric films. *Appl. Phys. A* **2016**, *122*, 802–810. [CrossRef]

41. Lee, H.U.; Lee, S.C.; Lee, Y.C.; Son, B.; Park, S.Y.; Lee, J.W.; Oh, Y.K.; Kim, Y.; Choi, S.; Lee, Y.S.; et al. Innovative three-dimensional (3D) eco-TiO_2 photocatalysts for practical environmental and bio-medical applications. *Sci. Rep.* **2014**, *4*, 6740. [CrossRef] [PubMed]

42. Hernández-Afonso, L.; Fernández-González, R.; Esparza, P.; Borges, M.E.; Dí González, S.; Canales-Vázquez, J.; Ruiz-Morales, J.C. Three dimensional printing of components and functional devices for energy and environmental applications. *Energy Environ. Sci.* **2017**, *10*, 846–859.

43. Giakoumaki, A.N.; Kenanakis, G.; Klini, A.; Androulidaki, M.; Viskadourakis, Z.; Farsari, M.; Selimis, A. 3D micro-structured arrays of ZnO nanorods. *Sci. Rep.* **2017**, *7*, 2100. [CrossRef]

44. Skorski, M.R.; Esenther, J.M.; Ahmed, Z.; Miller, A.E.; Hartings, M.R. The chemical, mechanical, and physical properties of 3D printed materials composed of TiO_2-ABS nanocomposites. *Sci. Technol. Adv. Mater.* **2016**, *17*, 89–97. [CrossRef] [PubMed]

45. Viskadourakis, Z.; Sevastaki, M.; Kenanakis, G. 3D structured nanocomposites by FDM process: A novel approach for large-scale photocatalytic applications. *Appl. Phys. A* **2018**, *124*, 585–593. [CrossRef]

46. Al-Kaf, A.G.; Naji, K.M.; Abdullah, Q.Y.M.; Edrees, W.H.A. Occurrence of Paracetamol in Aquatic Environments and Transformation by Microorganisms: A Review. *Chron. Pharm. Sci.* **2017**, *1*, 341–355.

47. Jones, O.A.H.; Voulvoulis, N.; Lester, J.N. The occurrence and removal of selected pharmaceutical compounds in a sewage treatment works utilizing activated sludge treatment. *Environ. Pollut.* **2007**, *145*, 738–744. [CrossRef]

48. Żur, J.; Piński, A.; Marchlewicz, A.; Hupert-Kocurek, K.; Wojcieszyńska, D.; Guzik, U. Organic micropollutants paracetamol and ibuprofen—toxicity, biodegradation, and genetic background of their utilization by bacteria. *Environ. Sci. Pollut. Res.* **2018**, *25*, 21498–21524. [CrossRef]

49. Sadeghzadeh-Attar, A.; Ghamsari, M.S.; Hajiesmaeilbaigi, F.; Mirdamadi, S.; Katagiri, K.; Koumoto, K. Modifier ligands effects on the synthesized TiO_2 nanocrystals. *J. Mater. Sci.* **2008**, *43*, 1723–1729. [CrossRef]

50. Bakardjieva, S.; Šubrt, J.; Štengl, V.; Dianez, M.J.; Sayagues, M.J. Photoactivity of Anatase-Rutile TiO_2 Nanocrystalline Mixtures Obtained by Heat Treatment of Homogeneously Precipitated Anatase. *Appl. Catal. B Environ.* **2005**, *58*, 193–202. [CrossRef]
51. Lubas, M.; Jasinski, J.J.; Sitarz, M.; Kurpaska, L.; Podsiad, P.; Jasinski, J. Raman spectroscopy of TiO_2 thin films formed by hybrid treatment for biomedical applications. *Spectrochim. Acta Part A* **2014**, *133*, 867–871. [CrossRef] [PubMed]
52. Shaikh, S.F.; Mane, R.S.; Min, B.K.; Hwang, Y.J.; Joo, O. D-sorbitol-induced phase control of TiO_2 nanoparticles and its application for dye-sensitized solar cells. *Sci. Rep.* **2016**, *6*, 20103. [CrossRef]
53. Shinde, D.V.; Patil, S.A.; Cho, K.; Ahn, D.Y.; Shrestha, N.K.; Mane, R.S.; Lee, J.K.; Han, S.H. Revisiting Metal Sulfide Semiconductors: A Solution-Based General Protocol for Thin Film Formation, Hall Effect Measurement, and Application Prospects. *Adv. Funct. Mater.* **2015**, *25*, 5739–5747. [CrossRef]
54. Sunada, K.; Kikuchi, Y.; Hashimoto, K.; Fujishima, A. Bactericidal and Detoxification Effects of TiO_2 Thin Film Photocatalysts. *Environ. Sci. Technol.* **1998**, *32*, 726–728. [CrossRef]
55. Rehman, S.; Ullah, R.; Butt, A.M.; Gohar, N.D. Strategies of making TiO_2 and ZnO visible light active. *J. Hazard. Mater.* **2009**, *170*, 560–569. [CrossRef]

Publisher's Note: MDPI stays neutral with regard to jurisdictional claims in published maps and institutional affiliations.

 *nanomaterials*

Article

Thickness Effect on Some Physical Properties of RF Sputtered ZnTe Thin Films for Potential Photovoltaic Applications

Dumitru Manica [1,2], **Vlad-Andrei Antohe** [1,3], **Antoniu Moldovan** [2], **Rovena Pascu** [2], **Sorina Iftimie** [1], **Lucian Ion** [1], **Mirela Petruta Suchea** [4,5,*] and **Ştefan Antohe** [1,6,*]

1 Faculty of Physics, University of Bucharest, 077125 Magurele, Romania; dumitru.manica@inflpr.ro (D.M.); vlad.antohe@fizica.unibuc.ro (V.-A.A.); sorina.iftimie@fizica.unibuc.ro (S.I.); lucian@solid.fizica.unibuc.ro (L.I.)
2 National Institute for Laser, Plasma and Radiation Physics, 077125 Magurele, Romania; antoniu.moldovan@inflpr.ro (A.M.); rovena.pascu@inflpr.ro (R.P.)
3 Institute of Condensed Matter and Nanosciences, "Université Catholique de Louvain", B-1348 Louvain-la-Neuve, Belgium
4 Center of Materials Technology and Photonics, School of Engineering, Hellenic Mediterranean University, 71410 Heraklion, Greece
5 National Institute for Research and Development in Microtechnologies, 023573 Bucharest, Romania
6 Academy of Romanian Scientists, 050044 Bucharest, Romania
* Correspondence: mirasuchea@hmu.gr or mira.suchea@imt.ro (M.P.S.); santohe@solid.fizica.unibuc.ro (Ş.A.)

Citation: Manica, D.; Antohe, V.-A.; Moldovan, A.; Pascu, R.; Iftimie, S.; Ion, L.; Suchea, M.P.; Antohe, Ş. Thickness Effect on Some Physical Properties of RF Sputtered ZnTe Thin Films for Potential Photovoltaic Applications. *Nanomaterials* **2021**, *11*, 2286. https://doi.org/10.3390/nano11092286

Academic Editor: Christian Mitterer

Received: 5 August 2021
Accepted: 31 August 2021
Published: 2 September 2021

Publisher's Note: MDPI stays neutral with regard to jurisdictional claims in published maps and institutional affiliations.

Abstract: Zinc telluride thin films with different thicknesses were grown onto glass substrates by the rf magnetron sputtering technique, using time as a variable growth parameter. All other deposition process parameters were kept constant. The deposited thin films with thickness from 75 to 460 nm were characterized using X-ray diffraction, electron microscopy, atomic force microscopy, ellipsometry, and UV-Vis spectroscopy, to evaluate their structures, surface morphology, topology, and optical properties. It was found out that the deposition time increase leads to a larger growth rate. This determines significant changes on the ZnTe thin film structures and their surface morphology. Characteristic surface metrology parameter values varied, and the surface texture evolved with the thickness increase. Optical bandgap energy values slightly decreased as the thickness increased, while the mean grains radius remained almost constant at ~9 nm, and the surface to volume ratio of the films decreased by two orders of magnitude. This study is the first (to our knowledge) that thoroughly considered the correlation of film thickness with ZnTe structuring and surface morphology characteristic parameters. It adds value to the existing knowledge regarding ZnTe thin film fabrication, for various applications in electronic and optoelectronic devices, including photovoltaics.

Keywords: ZnTe; thin films; rf-magneton sputtering; AFM; surface metrology; thickness effect

1. Introduction

There is a need for new nanostructured materials with enhanced properties, due to the development of new systems and materials that use nanotechnologies. Due to characteristics obtained at the time of the deposition process such as: low resistivity, high transparency in the visible spectrum, etc. [1–4], thin layers of zinc telluride (ZnTe) are used in various modern technologies, which are implemented in various micro- and nano-structured devices, such as light emitting diodes, solar cells, photodetectors, etc. [1–18]. Although it is a popular material, there are few publications and studies regarding its material engineering. ZnTe is a sensitive material in the green spectral region, with a bandgap of 2.26 eV and a low electronic affinity of 3.53 eV; it can be used as a p-type buffer material in hetero-junction solar cells based on CdTe [10–13,18]. It can be used as back contact material to CdTe-based solar cells [14] in a multilayer device. It is a precious material from an ecological point of view; it can be used as a replacement for the CdS

Nanomaterials **2021**, *11*, 2286. https://doi.org/10.3390/nano11092286   https://www.mdpi.com/journal/nanomaterials

window layer [15]; the "condition" is that the conductivity is n-type. For the deposition of thin layers [16–20], more physical and chemical growth paths are known, such as molecular beam epitaxy (MBE) [21], chemical bath deposition (CBD), electron-beam evaporation, thermal evaporation, magnetron sputtering [12], and electrodeposition [19–22]. The ZnTe thin layer properties depend directly on the deposition method, deposition conditions, and the growth direction imposed by the substrate when the substrate is crystalline [5–8,23]. Some studies have established the ideal deposition conditions for various deposition methods, to obtain ZnTe layers with the necessary physical properties, a well-defined morphology, and crystal structure, but there is no systematic reproducible technology available for high performance ZnTe coating fabrications [2–9,24,25]. In this context, studies regarding correlation for growth conditions/parameters with ZnTe structuring and morphology, in correlation with their physical properties, are still very important.

ZnTe thin films presented in this scientific report were prepared by rf magnetron sputtering, varying the deposition time that was related to the deposition rate, and finally determining the film thickness. Depending on the selected conditions, we sought to obtain an intermediate layer for the formation of junctions and the facilitation of electric charge transfer, which can be used directly in multilayer solar cells. In order to understand the ZnTe film structuring during growth, the investigation was centered on structure and surface morphology evolution, and the surface characteristic parameters correlation with material optical properties, and how it changed depending on the deposition parameters. For this scope, the following characterization methods were used: scanning electron microscopy (SEM), X-ray diffraction (XRD), atomic force microscopy (AFM), ellipsometry (spectroelipsometer-SE), and UV-Vis absorption spectroscopy. SEM was used for films thickness evaluation in the cross section while XRD was used to verify material crystallinity. AFM in non-contact mode was used to characterize the surface morphology and topology. A WVASE spectroelipsometer with variable incidence angles (60°, 65°, and 70°) was used for optical characterization. Optical models were generated by WVASE32 software; the parameters n (refractive index), k (extinction coefficient), and roughness were measured by mounting the ellipsometer parameters Psi (Ψ) and delta (Δ).

2. Experimental Procedures

2.1. Fabrication Technique

Magnetron sputtering is an easy-to-use deposition method. Due to the simplicity and easy handling of the installation, it allows good control of the deposition parameters. These parameters are enclosure pressure, substrate target distance, substrate temperature, applied power in the deposition process, deposition time, etc. ZnTe thin films were grown onto BK7 optical glass substrates (Heinz Herenz, Hamburg, DE6, Germany). Prior to any deposition process, the substrates were cleaned in acetone (Chim Reactiv S.R.L., Bucharest, BUC, Romania) and isopropyl alcohol (Chim Reactiv S.R.L., Bucharest, BUC, Romania) for 15 min, for each procedure, and then were rinsed in deionized water dried under nitrogen flow.

For all fabricated samples, the substrate-target distance, working pressure, substrate temperature, and applied power were maintained constant at 10 cm, 0.86 Pa, 250 °C, and 100 W, respectively, and the deposition time varied at 5, 10, 15, and 20 min. For ease of discussion, in the whole manuscript, the samples were denoted ZnTe1 (5 min), ZnTe2 (10 min), ZnTe3 (15 min), and ZnTe4 (20 min).

2.2. Characterization Methods

In order to characterize the obtained thin layers, various characterization techniques were involved. The characterization techniques used were SEM, AFM, UV-Vis, SE, and XRD.

2.2.1. Morphological and Structural Characterization

SEM was used to study the film cross sections and to estimate their thickness. While AFM was used for detailed characterization of film surfaces.

2.2.2. SEM Characterization

The SEM cross-section micrographs were obtained in high vacuum by using a Tescan Vega XMU-II electron microscope (Brno-Kohoutovice, B, Czech Republic) operating at 30 kV with a detector for secondary electrons.

2.2.3. AFM Characterization

The morphology of the surface was analyzed by AFM topography measurements using a XE100 AFM, from Park Systems (Suwon, Republic of Korea). The measurements were carried out in a noncontact mode, using silicon cantilevers (PPP-NCHR, Nanosensors, Neuchatel, Switzerland). The surface was scanned into various areas of the films with sizes of 5 μm × 5 μm and 2 μm × 2 μm. The surface roughness characteristic parameters were estimated using the AFM software while the surface topology specific parameters and texturing were evaluated by using Gwyddion version 2.49 (2017) open access specialized SPM software [26].

2.2.4. X-ray Diffraction

The structural features of fabricated ZnTe films were investigated by XRD, using a Bruker D8 Discover diffractometer from Brucker (Bruker Nano GmbH Am Studio 2D, 12489 Berlin, Germany) using CuKα = 1.54 Å radiation in Bragg–Brentano theta–theta geometry. The scattered X-ray photons from samples were recorded in the 2θ range of 20–70° with a scanning rate of 0.04°/s at room temperature.

2.2.5. Optical Characterization

Spectrophotometry (UV-Vis)

The optical properties of thin films depend on the structure, composition, and physical and chemical properties of the material. Using UV-Vis optical spectroscopy, information about the structure of energy levels and bands and photoconduction mechanisms can be obtained. UV-Vis optical spectroscopy transmission measurements were performed in the wavelength ranges of 300–1500 nm, at room temperature, using a Lambda 750 spectrophotometer from Perkin Elmer (Norwalk, CT, USA).

Ellipsometry

One of the widely used characterization methods was ellipsometry. For the optical characterization of ZnTe samples, a Spectro-ellipsometer WVASE, (Lincoln, NE, USA) was used, with variable angles of incidence (60°, 65°, and 70°), for optical characterization, having high accuracy and precision with a wide spectral range of 250–1700 nm. Optical models were generated by WVASE32 software; n, k parameters, and roughness were measured by fitting the Psi (Ψ) and delta (Δ) parameters.

3. Results and Discussions

3.1. SEM Microscopy Analysis—Film Thickness Evaluation

All samples were characterized by SEM microscopy as described above. The SEM analysis shows the effect of the deposition parameters on the material structuring. For example, Figure 1 presents representative images of layers grown under similar conditions, except for the variation of the deposition time. One can see from the SEM characterization how the structuring evolves when the deposition time changes. The ZnTe layers are compact with a relatively constant thickness. Some large asperities can be observed on the thinner film surface. As the film grows, the surface becomes cleaner and the layers seem to become more compact. Using the SEM analyses options, the local thickness of the ZnTe layers deposited by rf magnetron sputtering vas estimated. To evaluate film thicknesses, a mean value was calculated using local values, measured as shown in the examples presented in Figure 1, in different locations along the substrate lengths. Mean thickness values (calculated as the arithmetic average of various local thickness measured along the cross sections of each film) were estimated to be ZnTe1 (5 min) 75 nm, ZnTe2

(10 min) 154 nm, ZnTe3 (15 min) 251 nm, and ZnTe4 (20 min) 461 nm, respectively, with an error bar of $\pm$ 10% for each value.

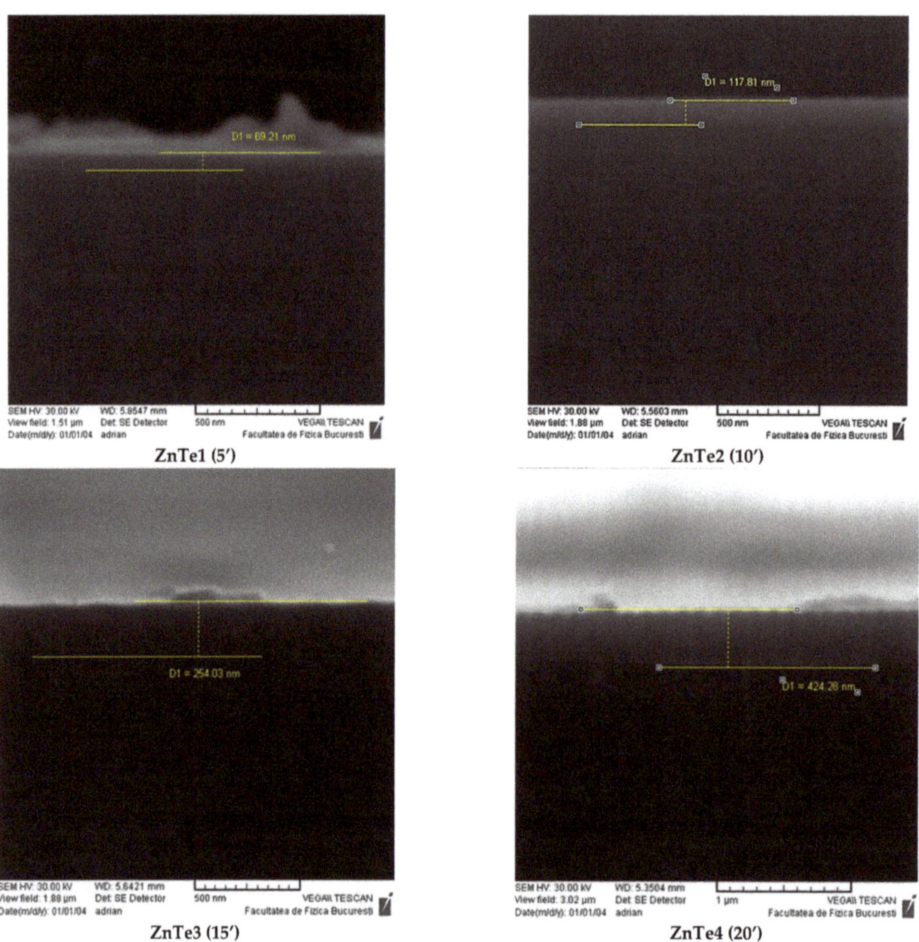

Figure 1. Examples of SEM images of cross sections of the fabricated ZnTe thin films (a local thickness estimation for each kind of film—the difference from the average values is due to local variations and it is in the errors limits).

It was noticed that, with the increase of the deposition time, the thickness of the thin layers also changed. It can be observed that the growth rate slightly increased from 15 nm/min for 5 min growth time; 15.4 nm/min for 10 min; 16.3 nm/min 15 min and becomes 23 nm/min for the thickest film. The increase of the growth rate in time can be attributed to the evolution of the growth mechanism of the films onto the substrate. In the early stage of nucleation, an island growth mechanism is present. This growth can be noticed when the adherence between the atom to atom is greater than the bonding between the substrate and the adatoms; it was observed on the thinnest films where the film was rough, island-like structured, and the thickness had larger variations across the substrate. At longer deposition times, the adatoms begin to accumulate; migration took place and the ZnTe layers with enhanced crystallinity were formed, as can be seen in Figure 1, and confirmed further by XRD analysis (see Section 3.2. Structural Characterization, X-ray Diffraction).

3.2. Structural Characterization, X-ray Diffraction

X-ray diffraction patterns were recorded and diffractograms for the samples were obtained at different deposition times: 5, 10, 15, and 20 min, labeled as ZnTe-5 min, ZnTe-10 min, ZnTe-15 min, and ZnTe-20 min, are presented in Figure 2. Diffraction features located at 25.24°, 42.34°, and 49.59° can be assigned unambiguously to (111), (220), and (311) reflections of the cubic ZnTe phase, according to the JCPDS database, card no. 01-0582. Therefore, ZnTe thicker films are polycrystalline with the respective zinc-blende structures. It can be observed that features are amorphous bands at 5 and 10 min, and evolve to diffraction peaks at 15 and 20 min. We should note that, at these stages, the chemical reaction of Zn with Te was complete, even if the crystal quality was poor—no Zn metallic phase or shift on the peak position can be observed. The evolution of the crystallinity at different stages of growth suggests the existence of a relationship between the crystal quality and the layer thickness, as a result of different deposition times. An increase of the crystallite size with the thickness for ZnTe films was also reported by Aboraia et al. [23], where films of different thicknesses were obtained by plasma immersion O^- ion implantation. The Scherrer equation was used so we could get a quantitative idea about the crystal quality. This equation relates the peak broadening by the crystal quality in the following way [24]:

$$\tau = \frac{k\lambda}{\beta cos\theta} \tag{1}$$

where k is the shape factor taken as 0.9, taking into account the spherical form of the grains, as shown in AFM images, λ = 0.154 nm is the wavelength of the X-rays and θ is the angular position. In the case of deposition times of 5 and 10 min, the diffraction features are amorphous and the Scherrer equation becomes inapplicable, at higher times (e.g., 15 and 20 min), the peak broadening on ZnTe (111) is 1.63° and 0.89°, respectively. Applying Equation (1), the mean crystallite size is 5.0 and 9.1 nm. At the same time, the position of the diffraction peaks remains unchanged, which indicates that the interplanar distances are preserved (e.g., d_{111} = 0.35 nm, d_{220} = 0.21 nm, and d_{311} = 0.18 nm according to Bragg's law) for different deposition times. By applying the standard relation between the interplanar distances and the lattice constant for cubic crystals [25], it was found that the unit cell parameter is 0.61 nm. As a conclusion, the XRD findings indicate that the different deposition times lead to different sizes for the crystalline domains for ZnTe, while the unit cell parameter remains unchanged. This can be further ascribed to a constant lattice strain at different stages of formation for ZnTe films.

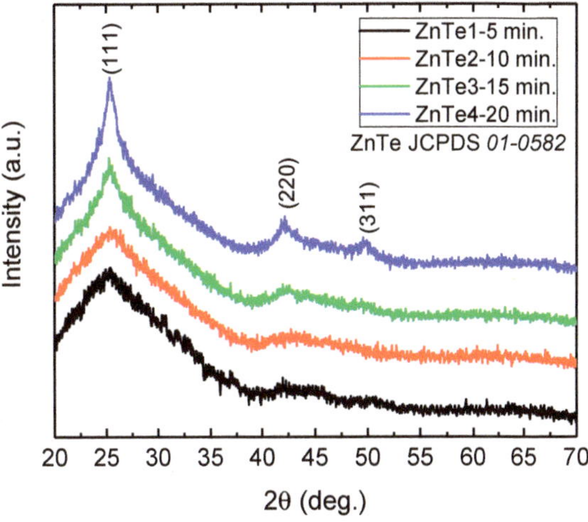

Figure 2. X-ray diffractograms of ZnTe films with different thicknesses.

3.3. AFM Characterization

AFM characterization of ZnTe thin films with different thicknesses show that all films have a granular structure that evolve as thickness increases. As can be observed from the two-dimensional (2D) 2 μm × 2 μm images presented in Figure 3, the thickness increase led to surfaces with a much smaller z-range fact, meaning surface flattening and smoothening. Using Gwyddion software, manual surface segmentation was performed to evaluate medium grain parameter sizes (http://gwyddion.net/download/user-guide/gwyddion-user-guide-en.pdf, accessed on 30 August 2021). Specific surface segmentations for each kind of ZnTe film surface are presented in Figure 3. In the figure, the violet regions represent water shade masking of deeper film regions where the grain segmentation could not be clearly performed. Grain borders assigned by segmentation are shown in a red–brownish color. The segmentation was performed to identify the largest number of similar-sized grains present on the 2 μm × 2 μm surface scan.

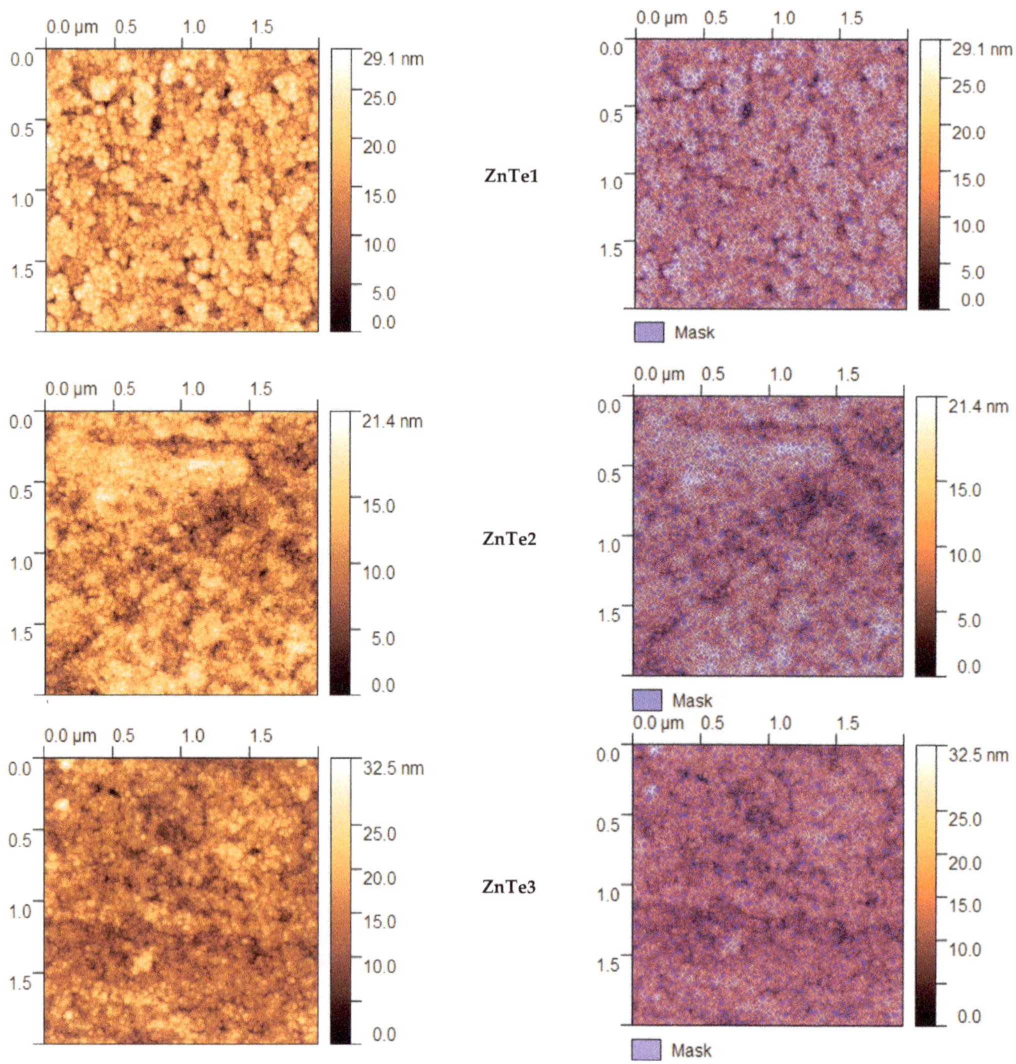

Figure 3. *Cont.*

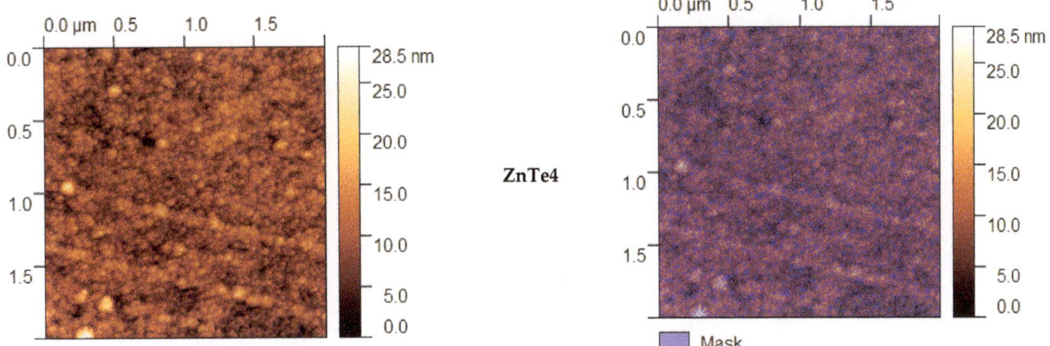

Figure 3. AFM characterization of ZnTe thin films with different thicknesses.

The calculated majoritarian specific parameter size grains are presented in Table 1. These were chosen based on statistical distributions presented in Figures 4–8. The calculus algorithms are open source, available at http://gwyddion.net/download/user-guide/gwyddion-user-guide-en.pdf, accessed on 30 August 2021.

Table 1. The size and specific parameters for the grains.

Sample	Mean Surface Area m^2	Number of Grains Used for the Surface Area Estimation	Mean Volume m^3	Number of Grains Used for Volume Estimation	Grain Boundaries nm	Number of Grains Used for Grain Boundaries Estimation	Mean Grain Radius nm	Number of Grains Used for Grain Radius Estimation	Surface Area/ Volume m^{-1}
ZnTe1	4.87E−16	6979	3.61E−26	7163	29.37	4851	9.46	5772	1.35E + 10
ZnTe2	4.42E−16	7270	2.26E−24	5773	31.36	5208	9.39	5884	1.95E + 8
ZnTe3	5.76E−16	7272	2.16E−25	7442	33.97	5283	9.95	6033	2.67E + 9
ZnTe4	4.52E−16	7337	2.39E−24	5432	31.59	5357	9.68	6210	1.89E + 8

Although the surface becomes flatter when the film thickens, it can be observed that the majoritarian grain on the surfaces remained the same, i.e., ~9–10 nm radius size for all of the films. The projected grain boundary values are also quite close—from 29 to 34 nm, while surface area and grain volumes are obviously more different due to strong surface texturing and a z-range drastic decrease. The surface-to-volume ratio is the amount of the surface area per unit volume of an object or collection of objects; in this case, the grains forming the film surface. It defines the relationship between the structures and functions in processes occurring through the surface and the volume of the film/layer. For the analyzed ZnTe thin films, the estimated surface to volume ratio was highest for the thinner film and was the lowest for the thicker. Additionally, the minimum circumcircle radius for each kind of grain detected on the surfaces, and the mean z-value were estimated for each above-presented AFM image, and are presented in the Figures 4–8. Table 1 presents the mean calculated values for the ZnTe films surface characteristic grains (specific surface area, grain volume, grain boundaries lengths, and surface to volume ratios) that can be further connected with the film growth mechanisms and physical properties. Based on Figures 4–8 that present the dimensional distribution of these surface characteristic parameter values for each of the films, the proper estimation of majoritarian grains is confirmed, and information regarding the other non-majoritarian grains (number and size ranges) present on the samples surfaces is provided. This offers the most complete kind of characterization regarding surface properties of the ZnTe films that one could obtain.

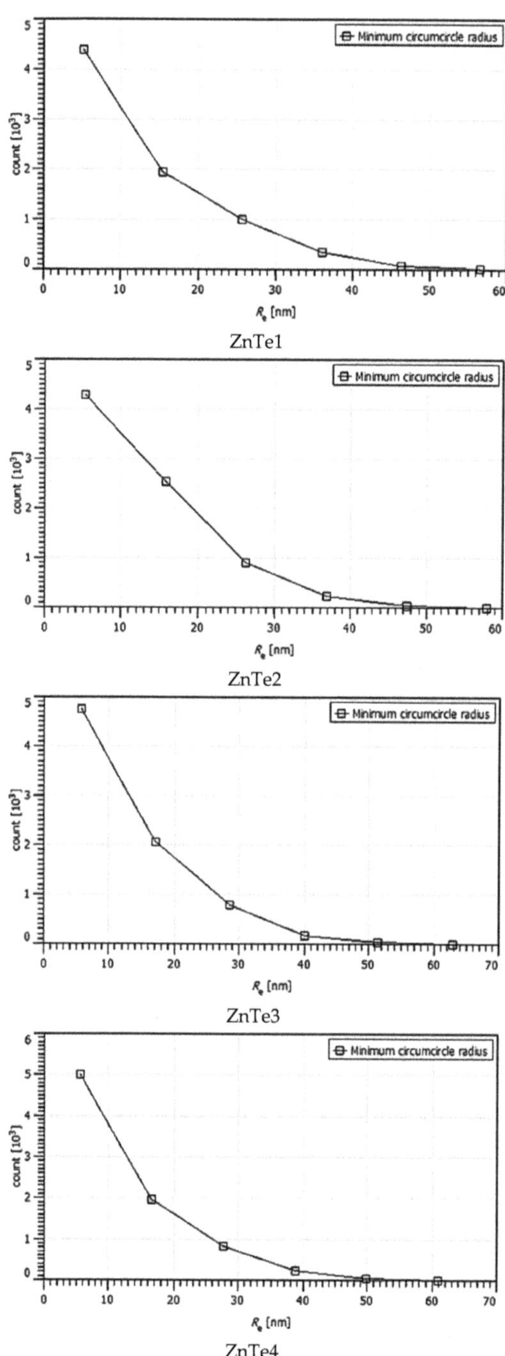

Figure 4. Minimum circumcircle radius value statistic variations onto surfaces of ZnTe1,2,3,4 thin films, presented in AFM images.

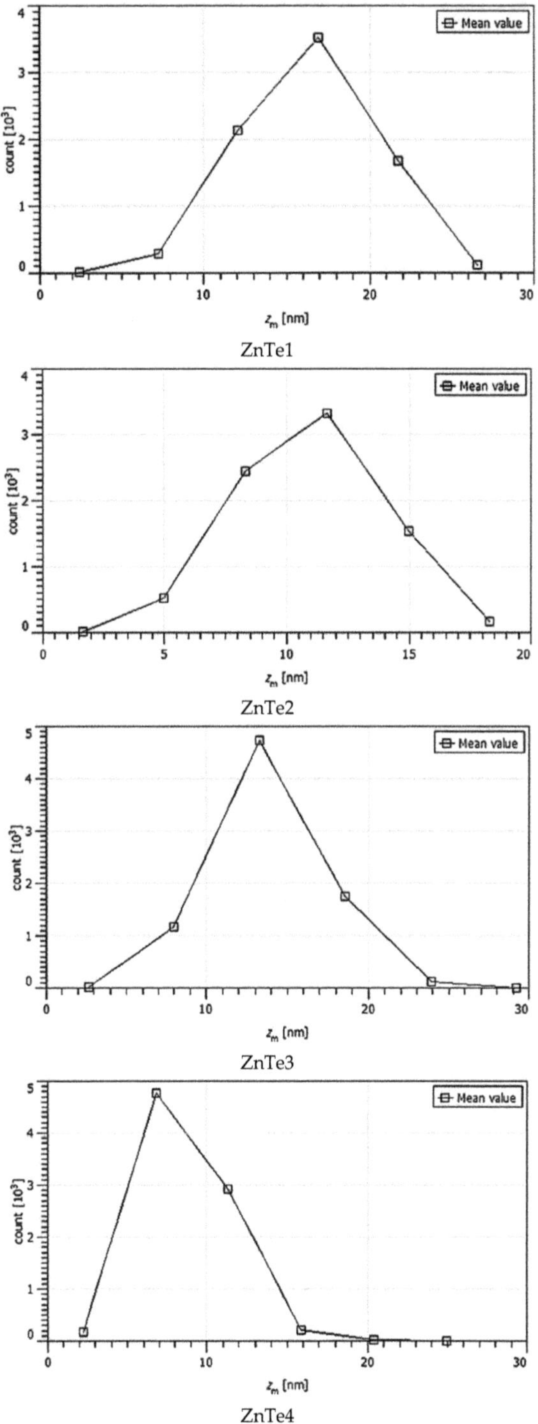

Figure 5. Mean value statistic variation onto surfaces of ZnTe1,2,3,4 thin films.

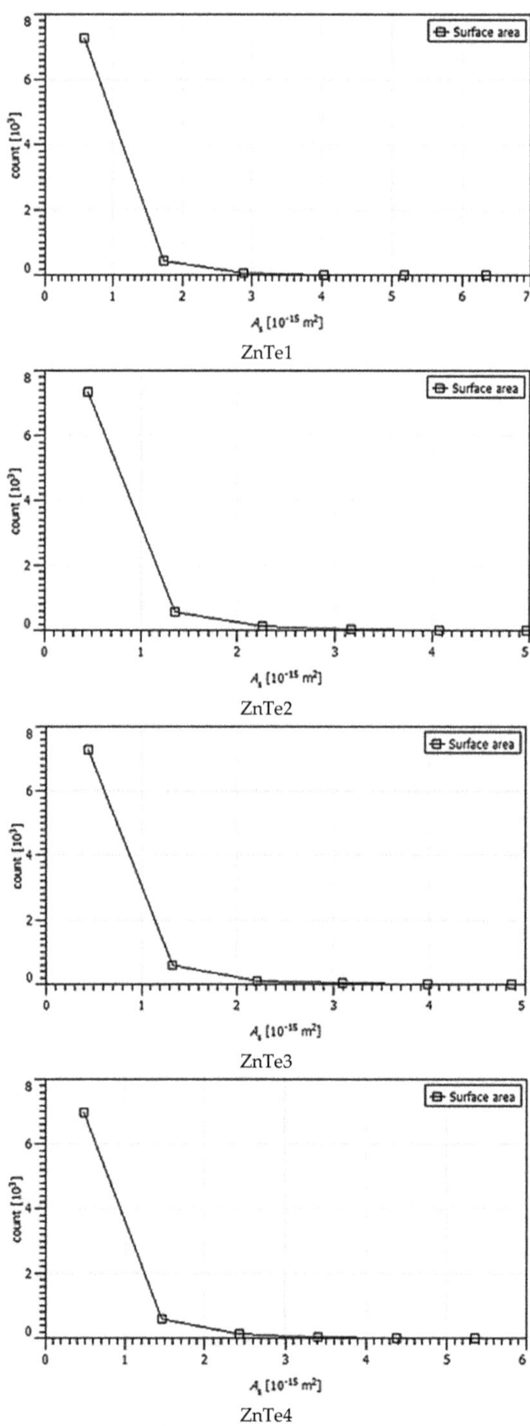

Figure 6. Surface area statistic variation of ZnTe1,2,3,4 thin films.

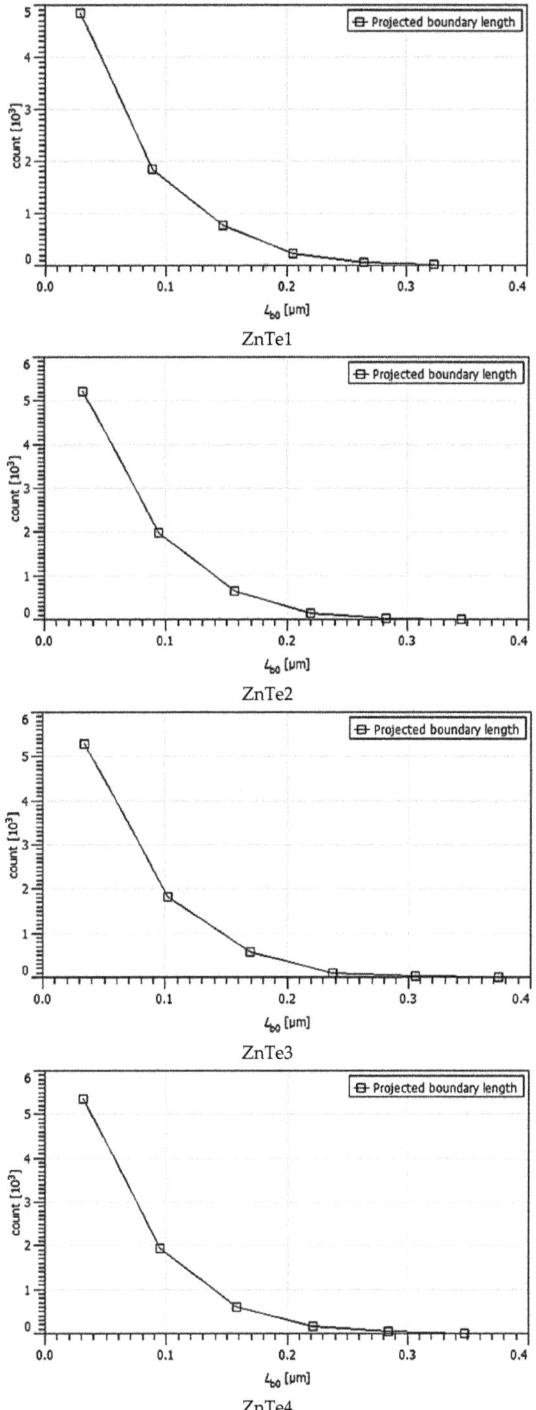

Figure 7. Projected boundary length statistic variation of ZnTe1,2,3,4 thin films.

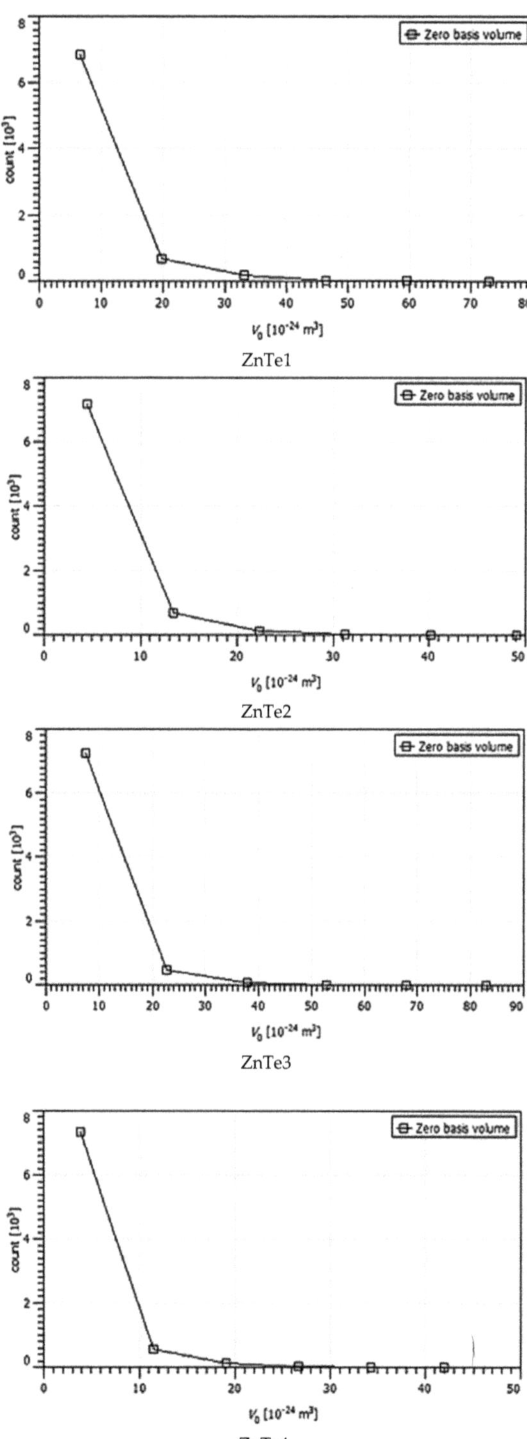

Figure 8. Zero basis volume statistic variation of ZnTe1,2,3,4 thin films.

The Figures 4–8 show the dimensional distributions of each parameter value onto the 2 μm × 2 μm surface scans (Figure 4).

To better analyze the surface roughness evolution, examples of three-dimensional (3D) AFM images of ZnTe films obtained from larger scan sizes (5 μm × 5 μm) are shown in Figure 9. The z-range of the specific 3D representations was chosen to be equal to the highest feature onto the surface, as a better way to observe the regularities and irregularities of the specific surface roughening process determined by film thickening.

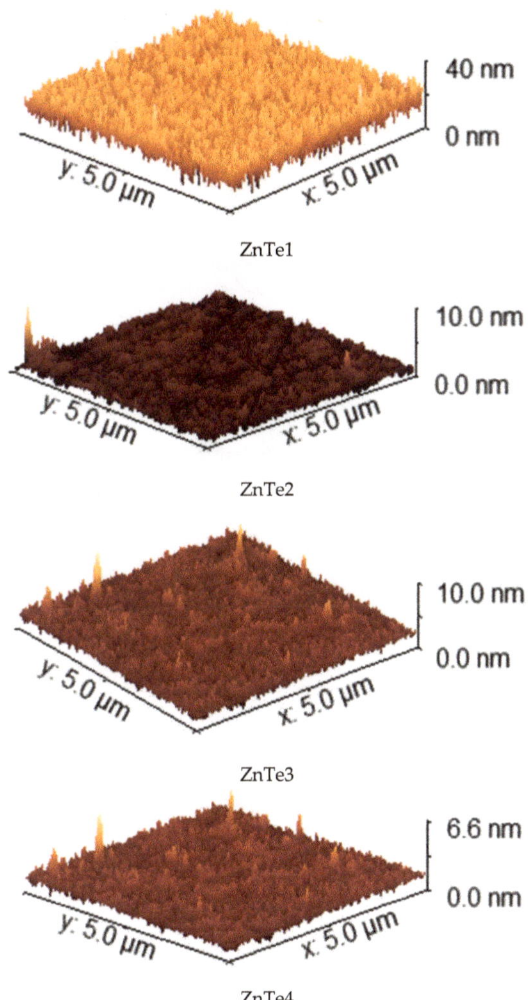

Figure 9. Examples of three-dimensional (3D) AFM images of ZnTe films.

It could be observed that the thicker films exhibit a waviness surface texture while the thinnest show pure grain distribution onto the surface. The growth time effect, onto surface structuring and morphology, is very strong. The thicker ZnTe film surfaces present valley regions, which become relatively smoother as the thickness increases. The cliff regions consist of spike structures that exhibit some orientation. The roughness parameters on surface morphology were estimated by analyzing the 5 μm × 5 μm topography images of the sample surface (Figure 9): average values, peaks, and valleys in the height direction

Ra, average amplitude in the height direction Rq, and average characteristics in the height direction Rsk.

The average roughness Ra involve peaks and valleys, the mean height measured in the entire area, useful for detecting the profile height sample characteristics. Ra variation typically signifies a change in the growth process. It can be observed that Ra decreases as the film thickness increases. Root mean square (RMS) roughness Rq is the square root of the standard distribution in the surface profile from the average height. Sq is more sensitive than the average roughness Sa for large deviations from the mean plan and is the most commonly reported measurement of surface roughness. ZnTe thin film thickness increase leads to an Rq decrease with ~25%. Ten-point mean height roughness (Rz) is a height parameter, the difference between the average of the five highest peaks and five lowest valleys in the sample surface Rz [26]. Due to the change in surface morphology from uniform granular distribution to a combination of isolated tall grains and tight packaging of grains at increasing thickness, one can see that Rz varies randomly in our case. Rsk values represent the degree of bias of the roughness shape (asperity). It can be observed that the Rq and Ra decrease severely with a film thickness increase, leading to precise determination of skewness Rsk and ten-point mean height roughness Rz, as can be seen in the high deviations in Table 2. The decrease of all surface morphology parameters indicates an inhomogeneity decrease. As the thickness increases, the average roughness Ra and RMS roughness Rq surface roughness decreases, restricting charged species to be adsorbed on the polycrystalline film. These changes may be related to the rf magnetron sputtering deposition process, which can stimulate the migration of grain boundaries and create more grains during the growth process. Moreover, at a high deposition rate, the supplementary energy encourages the atoms to acquire and occupy the correct site in the crystal lattice, such that the grains with lower surface energy will grow. These correlate well with the XRD observations and the fact that thinner films are amorphous while the thicker become crystalline. Further studies on intermediate growth times will be performed for a better solving of thin film growth mechanism evolution.

Table 2. Calculated roughness parameters on surface morphology.

	ZnTe1 (nm)	ZnTe 2 (nm)	ZnTe 3 (nm)	ZnTe 4 (nm)
Rq	4.08	3.45	3.29	3.09
Grain-wise Rpv	39.54	24.07	60.49	37.95
Ra	3.30	2.77	2.46	2.32
Rsk	0.18	0.04	−1.33	−0.92
Rz	36.32	23.52	54.33	36.03
Rku	2.87	2.79	14.55	6.13

3.4. Optical Characterization

3.4.1. Ellipsometry

Ellipsometry measurements were performed onto all ZnTe samples. Using the WVASE 32 software package, the simulation of the theoretical curves was performed. The obtained parameters were Ψ and Δ in the spectral range 250–1700 nm, scanning with a step of 2 nm. The analysis of the samples was performed at three incidence angles (60°, 65°, and 70°) with a step of 5° as described in references [27–30]. A ZnTe semiconductor model was selected according to the reference [30] for better measurement of bandwidth absorption, which is very important for the design of solar cells [29]. These aspects demonstrate the complexity of the thin film structures and influence of n, k parameters for many A2-B6 compounds [31–35]. As for the SE system, it was equipped with software for control and simulation of theoretical curves using a theoretical model [28]. The optical model was elaborated based on three layers: glass substrate, the ZnTe layer simulated using the Zinc telluride mathematical model in the WVASE database,, adjusting the thickness to fit the data, and the third layer—"srough". The zinc telluride mathematical model in the V VASE database is a GenOsc Tauc-Lorentz model. The model is fitted to minimize the mean square error (MSE) and reach a normal fit. Normal fit is reached by iterative approximations. The number of required iterations differs for each of the samples. Details regarding the simulation steps and models can be found in [27]. Film thickness and film roughness vary due to growth conditions. The scanning of the samples (ZnTe) and the substrate (BK7 Glass) were performed followed by the simulation of the theoretical curves with the help of the software, as can be seen from the examples presented in Figure 10.

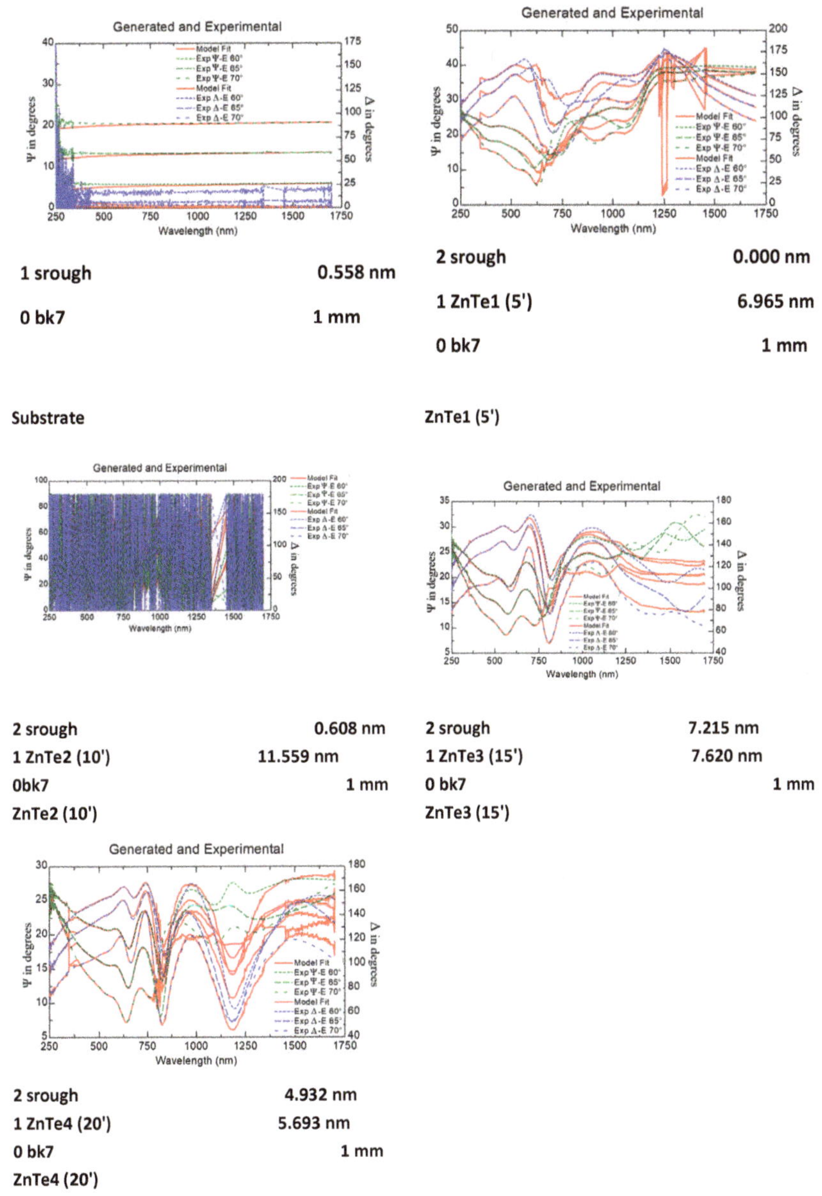

Figure 10. Theoretical and experimental ellipsometry data (Δ and Ψ) of ZnTe thin films at three incidence angles.

Figure 10 presents the SE measurements and theoretical modeling at three incidence angles (60°, 65°, and 70°) in the spectral range 250–1750 nm for estimation of optical roughness of the substrate and the ZnTe films analyzed during this study. Good qualitative agreement was found between the SE measurements and the theoretical simulations. The fitted SE spectra deviate significantly from the measured ones for the thinner films. These differences may be associated with the deviations of film stoichiometry and crystallinity in the various stages of growth from the ideal models. The differences are smaller for the thicker films. Further studies for better correlations are ongoing.

Table 3 represents a comparative presentation of the roughness measured by SE and RMS measured by AFM. The observed difference is generated by the different measurement principles of SE and AFM. SE measures the effect of roughness at the atomic size. The value of the RMS roughness is influenced from the peak to the height of the valley being approximately twice as high as the one estimated by SE measurements.

Table 3. Surface roughness measured by SE and AFM on ZnTe samples that have different deposition times.

Sample	Type of Substrate	Deposition Time (min)	Roughness Measured by SE (nm)	RMS Roughness AFM (nm)		Thickness Nonuniformity (%)	MSE
				$2 \times 2\ \mu m^2$	$5 \times 5\ \mu m^2$		
ZnTe1 (5′)	BK7	5	0.000	4.2	4.1	38.851%	15.01
ZnTe2 (10′)	BK7	10	0.608	3.1	3.5	63.064%	0.2747
ZnTe3 (15′)	BK7	15	7.215	3.4	3.3	100%	7.693
ZnTe4 (20′)	BK7	20	4.982	2.5	3	100%	19.55

3.4.2. Optical Spectroscopy and Bandgap

To study the optical properties of the fabricated samples, UV-Vis absorption measurements were performed. The UV-Vis absorption spectra of each sample, as well as the extrapolation used to calculate their optical bandgap using Tauc plots were obtained. Since ZnTe is a direct bandgap semiconductor, $(\alpha h v)^2$ was used to calculate the bandgaps. Representative examples of some absorption spectra and Tauc plots derived from the ZnTe thin film analysis are presented in Figure 11.

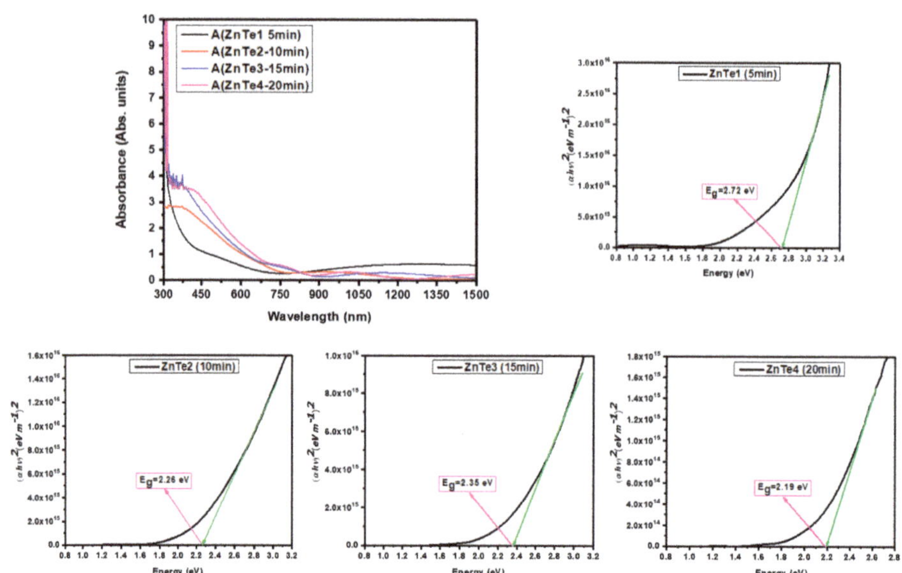

Figure 11. Examples of UV-Vis absorption spectra and of optical bandgap energy estimation using Tauc Plots.

Bandgap energy (Eg) values obtained for the optimized materials are presented in Table 4.

Table 4. Bandgap values.

Sample	Bandgap Eg(eV)
ZnTe1	2.72
ZnTe2	2.26
ZnTe3	2.35
ZnTe4	2.19

It was observed that the increased thickness leads to a decrease of optical transmittance in the visible region and the presence of the fringes in the NIR region of the electromagnetic spectrum (see Figure 12). The calculated bandgap values slightly decrease with the thickness increase.

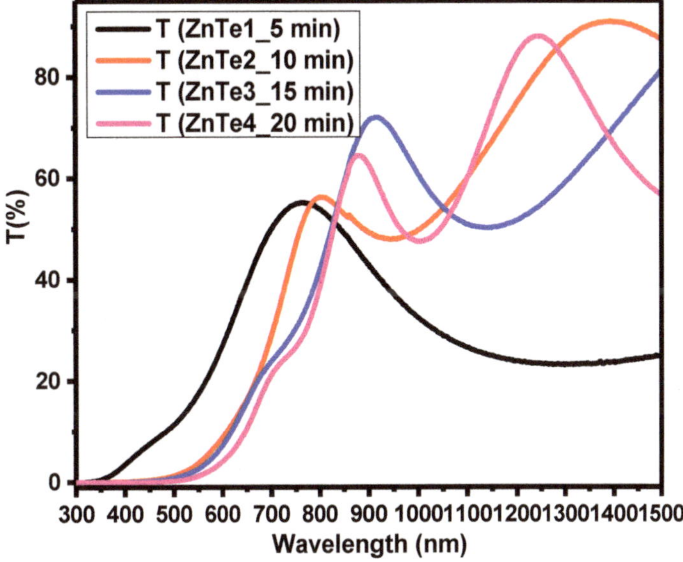

Figure 12. UV-Vis transmission spectroscopy spectra of ZnTe thin films with different thicknesses.

4. Conclusions and Perspectives

ZnTe thin films with different thicknesses on BK7 glass substrates were grown by the rf magnetron sputtering technique, using time as a variable growth parameter. All other deposition process parameters were kept constant. The fabricated thin films with mean thickness, ranging from 75 to 461 nm, were characterized using electron microscopy, X-rays diffraction, atomic force microscopy, ellipsometry, and UV-Vis spectroscopy, to evaluate their structures, surface morphology, topology, and optical properties. By using SEM to measure the obtained film thickness, it was found that the deposition time increase leads to a larger growth rate. This determines significant changes on ZnTe thin film structuring and surface morphology. Characteristic surface metrology parameter values vary, and surface texturing evolves with thickness increase, correlating well with XRD analysis findings that thinner films are amorphous, and at least 10 min of growth is needed for the crystalline material to start to form. Larger thickness films show diffraction peaks located at 25.24°, 42.34°, and 49.59°, which corresponds unambiguously to (111), (220), and (311) reflections of the cubic ZnTe phase, according to the JCPDS database with card no. 01-0582. For these films, the mean crystallite sizes are ~5 and ~9 nm, smaller than the surface grain size, estimated from the AFM analysis to be ~9 nm radius, i.e., ~18 nm diameter. The crystallite size shows values comparable to the ZnTe exciton radius estimated at ~6–7 nm [36], fact that opens interesting supplementary research perspective on size quantum effects on optical properties of such ZnTe films grown onto various substrates. Optical bandgap energy values slightly decrease as the thickness increases, while the mean grain radius remains almost constant at ~9 nm and the surface to volume ratio per grain decreases by two orders of magnitude. To our knowledge, this study is the first attempt that thoroughly considers the correlation of film thickness with ZnTe structuring and surface morphology characteristic parameters. It adds value to the existing knowledge on ZnTe thin film fabrication and their physical properties, tailored for various applications, including photovoltaics. There are ongoing studies regarding the correlation between structure and surface morphology, with the observed optical properties; this will lead to future scientific reports.

Author Contributions: Conceptualization, Ş.A.; methodology, Ş.A.; formal analysis, L.I., V.-A.A., S.I., M.P.S.; investigation, D.M., V.-A.A., S.I., A.M., R.P. and Ş.A.; resources, V.-A.A., S.I., M.P.S.; writing–original draft preparation, D.M., S.I.; M.P.S. and V.-A.A.; writing—review and editing, M.P.S. and Ş.A.; supervised and coordinated the experimental work. All authors have read and agreed to the published version of the manuscript.

Funding: This research was funded by the "Executive Unit for Financing Higher Education, Research, Development and Innovation" (UEFISCDI, Romania), through grants: PN-III-P1-1.1-TE-2019-0868 (TE 115/2020) and PN-III-P1-1.1-TE-2019-0846 (TE 25/2020). M.P.S. contribution was partially financed by the Romanian Ministry of Research, Innovation and Digitalisation thorough "MICRO-NANO-SIS PLUS" core Programme.

Institutional Review Board Statement: Not applicable.

Informed Consent Statement: Not applicable.

Data Availability Statement: The raw and processed data required to reproduce these findings cannot be shared at this time due to technical or time limitations. The raw and processed data will be provided upon reasonable request to anyone interested anytime, until the technical problems will be solved.

Acknowledgments: D.M. thanks Cosmin Romanitan from IMT Bucharest for valuable advice on XRD data analysis.

Conflicts of Interest: The authors declare no conflict of interest. The funders had no role in the design of the study; in the collection, analyses, or interpretation of data; in the writing of the manuscript, or in the decision to publish the results.

References

1. Rusu, G.I.; Prepeliţă, P.; Rusu, R.S.; Apetroaie, N.; Oniciuc, G.; Amariei, A. On the structural and optical characteristics of zinc telluride thin films. *J. Optoelectron. Adv. Mater.* **2006**, *8*, 922–926.
2. Rathod, J.R.; Patel, H.S.; Patel, K.D.; Pathak, V.M. Structural and Optical Characterization of Zinc Telluride Thin Films. *Adv. Mater. Res.* **2013**, *665*, 254–262. [CrossRef]
3. Shanmugan, S.; Mutharasu, D. Optical Properties and Surface Morphology of Zinc Telluride Thin Films Prepared by Stacked Elemental Layer Method. *Mater. Sci.* **2012**, *18*, 107–111. [CrossRef]
4. Sun, J.; Shen, Y.; Chen, R.; Huang, J.; Cao, M.; Gu, F.; Wang, L.; Min, J. Effect of ZnTe transition layer to the performance of CdZnTe/GaN multilayer films for solar-blind photodetector applications. *J. Phys. D Appl. Phys.* **2020**, *53*, 415105. [CrossRef]
5. Suthar, D.; Himanshu; Patel, S.L.; Chander, S.; Kannan, M.D.; Dhaka, M.S. Enhanced physicochemical properties of ZnTe thin films as potential buffer layer in solar cell applications. *Solid State Sci.* **2020**, *107*, 106346. [CrossRef]
6. Zarei, R.; Ehsani, M.H.; Dizaji, H.R. An investigation on structural and optical properties of nanocolumnar ZnTe thin films grown by glancing angle technique. *Mater. Res. Express* **2020**, *7*, 026419. [CrossRef]
7. Isik, M.; Gullu, H.H.; Parlak, M.; Gasanly, N.M. Synthesis and temperature-tuned band gap characteristics of magnetron sputtered ZnTe thin films. *Phys. B Condens. Matter* **2020**, *582*, 411968. [CrossRef]
8. Toma, O.; Ion, L.; Iftimie, S.; Antohe, V.A.; Radu, A.; Raduta, A.M.; Manica, D.; Antohe, S. Physical properties of rf-sputtered ZnS and ZnSe thin films used for doubleheterojunction ZnS/ZnSe/CdTe photovoltaic structures. *Appl. Surf. Sci.* **2019**, *478*, 831–839. [CrossRef]
9. Maki, S.A.; Hanan, H.K. The Structural and Optical Properties of Zinc Telluride Thin Films by Vacuum Thermal Evaporation Technique. *Ibn AL-Haitham J. Pure Appl. Sci.* **2017**, *29*, 70–80.

10. Ikhioya, I.L. Optical and electrical properties of znte thin films using electrodeposition technique. *Int. J. Innov. Appl. Stud.* **2015**, *12*, 369–373.

11. Li, J.; Diercks, D.R.; Ohno, T.R.; Warren, C.W.; Lonergan, M.C.; Beach, J.D.; Wolden, C.A. Controlled activation of ZnTe: Cu contacted CdTe solar cells using rapid thermal processing. *Sol. Energy Mater. Sol. Cells* **2015**, *133*, 208–215. [CrossRef]

12. Gessert, T.A.; Asher, S.; Johnston, S.; Young, M.; Dippo, P.; Corwine, C. Analysis of CdS/CdTe devices incorporating a ZnTe:Cu/Ti Contact. *Thin Solid Film.* **2007**, *515*, 6103–6106. [CrossRef]

13. Lopez, J.D.; Tirado-Mejia, L.; Ariza-Calderon, H.; Riascos, H.; de Anda, F.; Mosquera, E. Structural and optical properties of gadolinium doped ZnTe thin films. *Mater. Lett.* **2020**, *268*, 127562. [CrossRef]

14. Bhahada, K.C.; Tripathi, B.; Acharya, N.K.; Kulriya, P.K.; Vijay, Y.K. Formation of ZnTe by stacked elemental layer method. *Appl. Surf. Sci.* **2008**, *255*, 2143–2148. [CrossRef]

15. Pfisterer, F.; Schock, H.W. ZnTe-CdS Thin-Film photo-voltaic cells. *J. Cryst. Growth* **1982**, *59*, 432–439. [CrossRef]

16. Naifar, A.; Zeiri, N.; Nasrallah, S.A.-B.; Said, M. Optical properties of CdSe/ZnTe type II core shell nanostructures. *Optik* **2017**, *146*, 90–97. [CrossRef]

17. Mahmood, W.; Shah, N.A. Effects of metal doping on the physical properties of ZnTe thin films. *Curr. Appl. Phys.* **2014**, *14*, 282–286. [CrossRef]

18. Wolden, C.A.; Abbas, A.; Li, J.; Diercks, D.R.; Meysing, D.M.; Ohno, T.R.; Beach, J.D.; Barnes, T.M.; Walls, J.M. The roles of ZnTe buffer layers on CdTe solar cell performance. *Sol. Energy Mater. Sol. Cells* **2016**, *147*, 203–210. [CrossRef]

19. Ogawa, H.; Irfan, G.S.; Nakayama, H.; Nishio, M.; Yoshida, A. Growth of low resistivity n-type ZnTe by metalorganic vapor phase epitaxy. *Jpn. J. Appl. Phys.* **1994**, *33*, L980–L982. [CrossRef]

20. Rehman, K.M.U.; Liu, X.S.; Riaz, M.; Yang, Y.J.; Feng, S.J.; Khan, M.W.; Ahmad, A.; Shezad, M.; Wazir, Z.; Ali, Z.; et al. Fabrication and characterization of Zinc Telluride (ZnTe) thin films grown on glass substrates. *Phys. B Condens. Matter* **2019**, *560*, 204–207. [CrossRef]

21. Chang, J.H.; Takai, T.; Koo, B.H.; Song, J.S.; Handa, T.; Yao, T. Aluminum-doped n-type ZnTe layers grown by molecular-beam epitaxy. *Appl. Phys. Lett.* **2001**, *79*, 785–787. [CrossRef]

22. Mahalingam, T.; John, V.S.; Rajendran, S.; Sebastian, P.J. Electrochemical deposition of ZnTe thin films. *Semicond. Sci. Technol.* **2002**, *17*, 465–470. [CrossRef]

23. Structural and Optical Properties of ZnTe Thin Films Induced by Plasma Immersion O−ion Implantation. Available online: https://www.semanticscholar.org/paper/Structural-and-optical-properties-of-ZnTe-thin-by-O-Aboraia-Ahmad/b4307edf5d7445fb79363dbe93d32cd0ac20a88a (accessed on 30 August 2021).

24. Patterson, A.L. The Scherrer Formula for X-ray Particle Size Determination. *Phys. Rev.* **1939**, *56*, 978–982. [CrossRef]

25. Waseda, Y.; Matsubara, E.; Shinoda, K. *X-ray Diffraction Crystallography: Introduction, Examples and Solved Problems*; Springer: Berlin/Heidelberg, Germany, 2011. [CrossRef]

26. Available online: http://gwyddion.net/documentation/user-guide-en/roughness-iso.html (accessed on 30 August 2021).

27. J.A. Woollam Co., Inc. VASE®Specifications—Technical Specifications. 2007. Available online: https://www.jawoollam.com/download/pdfs/vase-brochure.pdf (accessed on 30 August 2021).

28. Fujiwara, H. *Spectroscopic Ellipsometry: Principles and Applications*; John Wiley Sons: Hoboken, NJ, USA, 2007.

29. Potlog, T.; Maticiuc, N.; Mirzac, A.; Dumitriu, P.; Scortescu, D. Structural and optical properties of ZnTe thin films. In Proceedings of the CAS 2012 (International Semiconductor Conference), Sinaia, Romania, 15–17 October 2012; pp. 321–324. [CrossRef]

30. Toma, O.; Pascu, R.; Dinescu, M.; Besleaga, C.; Mitran, T.L.; Scarisoreanu, N.; Antohe, S. Growth and Characterization of Nanocrystalline CdS Thin Films. *Chalcogenide Lett.* **2011**, *8*, 541–548.

31. Antohe, S.; Ion, L.; Girtan, M.; Toma, O. Optical and morphological studies of thermally vacuum evaporated ZnSe thin films. *Rom. Rep. Phys.* **2013**, *65*, 805–811.

32. Ion, L.; Iftimie, S.; Radu, A.; Antohe, V.A.; Toma, O.; Antohe, S. Physical properties of RF-sputtered ZnSe thin films for photovoltaic applications: Influence of film thickness. *Proc. Rom. Acad. Ser. A* **2021**, *22*, 25–34.

33. Toma, O.; Ion, L.; Iftimie, S.; Radu, A.; Antohe, S. Structural, morphological and optical properties of rf-Sputtered CdS thin films. *Mater. Des.* **2016**, *100*, 198–203. [CrossRef]

34. Toma, O.; Ion, L.; Girtan, M.; Antohe, S. Optical, morphological and electrical studies of thermally vacuum evaporated CdTe thin films for photovoltaic applications. *Sol. Energy* **2014**, *108*, 51–60. [CrossRef]

35. Ruxandra, V.; Antohe, S. The effect of the electron irradiation on the electrical properties of thin polycrystalline CdS layers. *J. Appl. Phys.* **1998**, *84*, 727–733. [CrossRef]

36. Yoffe, A.D. Low-dimensional systems: Quantum size effects and electronic properties of semiconductor microcrystallites (zero-dimensional systems) and some quasi-two-dimensional systems. *Adv. Phys.* **1993**, *42*, 173–262. [CrossRef]

Article

From Chip Size to Wafer-Scale Nanoporous Gold Reliable Fabrication Using Low Currents Electrochemical Etching

Pericle Varasteanu [1,2,*], Cosmin Romanitan [1], Alexandru Bujor [1,3], Oana Tutunaru [1], Gabriel Craciun [1], Iuliana Mihalache [1], Antonio Radoi [1] and Mihaela Kusko [1,*]

[1] National Institute for Research and Development in Microtechnology (IMT-Bucharest), 126A Erou Iancu Nicolae Street, 077190 Voluntari, Romania; cosmin.romanitan@imt.ro (C.R.); alexandru.bujor@imt.ro (A.B.); oana.tutunaru@imt.ro (O.T.); gabriel.craciun@imt.ro (G.C.); iuliana.mihalache@imt.ro (I.M.); antonio.radoi@imt.ro (A.R.)

[2] Faculty of Physics, University of Bucharest, 405 Atomistilor Street, 077125 Magurele, Romania

[3] Faculty of Chemistry, University of Bucharest, 90-92 Panduri Street, 050663 Bucharest, Romania

* Correspondence: pericle.varasteanu@imt.ro (P.V.); mihaela.kusko@imt.ro (M.K.); Tel.: +40-21-269-07-68 (P.V.)

Received: 30 October 2020; Accepted: 16 November 2020; Published: 23 November 2020

Abstract: We report a simple, scalable route to wafer-size processing for fabrication of tunable nanoporous gold (NPG) by the anodization process at low constant current in a solution of hydrofluoric acid and dimethylformamide. Microstructural, optical, and electrochemical investigations were employed for a systematic analysis of the sample porosity evolution while increasing the anodization duration, namely the small angle X-ray scattering (SAXS) technique and electrochemical impedance spectroscopy (EIS). Whereas the SAXS analysis practically completes the scanning electronic microscopy (SEM) investigations and provides data about the impact of the etching time on the nanoporous gold layers in terms of fractal dimension and average pore surface area, the EIS analysis was used to estimate the electroactive area, the associated roughness factor, as well as the heterogeneous electron transfer rate constant. The bridge between the analyses is made by the scanning electrochemical microscopy (SECM) survey, which practically correlates the surface morphology with the electrochemical activity. The results were correlated to endorse the control over the gold film nanostructuration process deposited directly on the substrate that can be further subjected to different technological processes, retaining its properties. The results show that the anodization duration influences the surface area, which subsequently modifies the properties of NPG, thus enabling tuning the samples for specific applications, either optical or chemical.

Keywords: nanoporous gold; large-scale fabrication; absorbance; X-ray diffraction; electrochemical impedance spectroscopy; scanning electrochemical microscopy

---

1. Introduction

Nanoporous gold (NPG) has been known since the ancient times; the South American population and eastern part of Europe throughout the centuries used to etch (dealloy) and polish the surface of gold–copper alloys to create the illusion of bulk gold of shininess, a process known as depletion gilding or gold colouration (*mise-en-couleur*) [1]. The percolating structure of NPG, consisting of a matrix of interconnected gold ligaments and voids, was not known until 1960, when transmission electron microscopy was employed to study the corrosion of gold (Au) alloys [2].

Therefore, NPG preserves the properties of gold, such as high conductivity, good chemical stability, as well as biocompatibility, but also presents significant advantages, mainly determined by the large surface to volume ratio. Consequently, it becomes a good candidate for improving next generation chemical/biochemical sensors [3–5]. Thus, immunosensors, enzyme-based biosensors, or DNA sensors are only a few examples of sensors where an increase from 2 to 1000 times of the surface area in comparison with the planar gold counterpart of an equivalent geometric area would improve sensitivity and lower detection limits [6] Besides, Faradaic current scales linearly with the electrode area, which means that the NPG electrodes outperform the flat gold electrodes in terms of the amount of reagents immobilized on surface, higher signal to noise ratio, and smaller impedance [7]. In addition, NPG electrodes have recently been introduced as good current collectors in supercapacitors, firstly because they provide a large surface area to deposit active materials and guarantee effective charge transport [8], but also because they enhance the specific capacitance of poor-conductive active materials [7,9]. As for plasmonic sensing, NPG provides simple excitation schemes of strong plasmonic resonances localized on the nanoporous structure ligaments, with the resulting plasmon-induced electric field being the primary mechanism in surface-enhanced spectroscopy, including surface-enhanced Raman scattering (SERS), surface-enhanced infrared absorption (SEIRA), and surface-enhanced fluorescence (SEF) [10]. Furthermore, it was shown recently that NPG could be used as a high performance plasmonic biosensor in the near infrared spectral region (NIR), where the flat gold platforms exhibit low performances [10,11]. The catalytic activity is originated by the molecular oxygen activation, where the low coordinated gold atoms increase the strength of the oxygen binding [12]. Thus, high catalytic activity was reported towards CO oxidation of [13,14], but also hydrogen oxidation [15] or selective oxidation of alcohols [4]. An intriguing feature of NPG is the catalytic activity without additional supporting transition metal oxides nanoparticles [13]. Furthermore, taking into account that the upper limit of gold nanoparticles that could sustain catalytic activity is 5 nm [16], the ligament size larger than 10 nm seems to be too large for catalytic performances, but it is successfully compensated by the presence of steps, kinks, surface defects, twin boundaries, and dislocations in the NPG architecture [17–19].

Even though the gold nanostructures for various applications could be fabricated using lithographic techniques, the expensive equipment, processing cost, and small nanostructured area have slowed down the development of these structures. Moreover, it was shown that the NPG structures exhibit the same performances as the structures obtained from electron beam lithography [20], which means that the higher cost of fabrication and small area of the EBL fabricated gold nanostructures could be overcome by the NPG in several applications. The most used methods to fabricate NPG are dealloying [21] and templating [22]. In general, a binary gold alloys with different atomic weights are employed in dealloying, where, by chemical or electrochemical corrosion, the less noble metal is selectively removed from the compound, leading to a "sponge-like" structure with voids and ligaments [4,17,23]. Porosity depends on the atomic weight ratio in the alloy, thickness, and corrosion time. Templating represents deposition of the gold over an artificial or natural matrix with different sizes, with the porous structure being obtained after removing the template [24]. Several drawbacks are associated with these techniques, such as the increased stress within the structure, which might lead to cracks that can be detrimental especially for further technological processing or sensing applications [25], or the residual less noble metal after dealloying that artificially increases the double layer capacitance of the porous film [26,27]. Last, but not the least, the technology is problematic when the fragile, difficult to handle dealloyed leaves are then transferred onto a substrate (e.g., silicon wafers) that would be further subjected to the different photolithographic processes. Deposition of the alloy film directly by sputtering on the processable substrate is an alternative, but it significantly increases the fabrication cost. An attempt to obtain NPG directly from thin metallic gold films was realized using electrochemical porosification under high anodic bias up to 40 V [3,28]. However, a very high-bias regime induces local heating effects that focus charge carriers to the pore tips, considerably enlarging their diameters and imposing a limited process duration. Herein, based on our previous experience in fabricating porous silicon [29], a porosification procedure for Au was established using a constant applied current that can be generated with maximum 2.5 V bias, with the process being monitored after different periods of time, from 200 to 1000 s, when, lately, the entire gold film was deeply nanostructured.

The obtained NPG films were characterized by means of scanning electron microscopy and correlated with the X-ray diffraction studies in order to obtain morpho-structural information (i.e., porosity, ligament diameter, and surface area) and to propose a mechanism of porosification under constant current anodization. Electrochemical investigations were further conducted to examine and visualize the active surface area in comparison with the flat gold. Not only does the process allow a good control over the NPG morpho-structural, optical, and electrochemical properties, it is also a fast, scalable, and cost-effective method, which is always desirable when talking about device manufacturing.

2. Materials and Methods

The NPG samples were prepared using as a starting substrate 4 inch silicon wafers (SIEGERT WAFER GmbH, Aachen, Germany) deposited with 20 nm chromium as an adherence layer and 300 nm gold by DC magnetron sputtering. Autolab PGCSTAT302n (Metrohm, Utrecht, The Netherlands) was used for NPG samples' fabrication and for electrochemical analysis. A two-electrode configuration was employed, where the gold layer was used as a working electrode and round platinum grid with the diameter equal to the exposed area of the cell (0.5 cm^2) as a counter electrode. Gold electrode was placed horizontally in the cell (Figure S1), while the platinum circular electrode was placed parallel to the surface at a height of approximately 2 mm away. The electrolyte contains equal volumes of 48% aqueous hydrofluoric acid (HF, analytical grade) and organic solvent dimethylformamide (DMF, 99.5% analytical grade). The volume of etching solution was 1 mL. The NPG samples were obtained using a 15 mA anodization constant current, and varying the process duration from 200 to 1000 s. The samples were labelled in the following manner: NPG200, NPG400, and so on, where the associated numbers are consistent with the process duration (Figure 1).

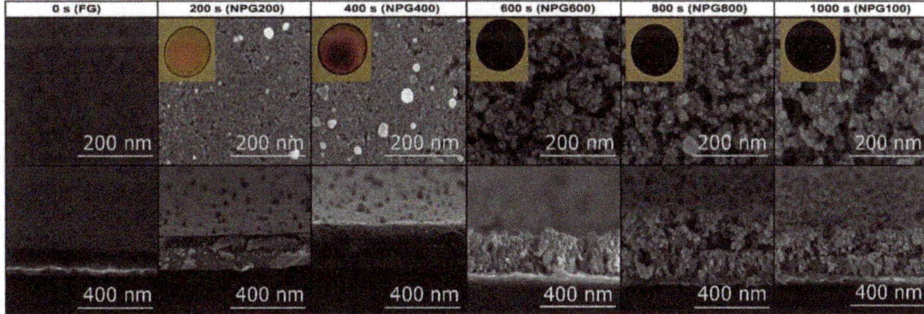

Figure 1. Top and cross-section scanning electron microscopy (SEM) micrographs of nanoporous gold (NPG) samples obtained after different etching times. Inset: photograph images of the experimental samples.

Scanning electron microscopy (SEM) micrographs were obtained using field emission gun scanning electron microscope (FEI Nova NanoSEM 630, Hillsboro, OR, USA). Absorption spectra were measured using integrating sphere in the wavelength range of 350–800 nm with a combined time resolved and steady state fluorescence spectrometer (Edinburgh Instruments, Livingston, UK). For the optical measurements, a black tape with a circular opening with the area of 0.5 cm^2 where the porous layer is positioned was used in order to avoid the unwanted reflections from the remaining gold surface. The microstructure of the obtained samples, as well as the surface/pore morphology, were analyzed using a 9 kW Rigaku SmartLab diffractometer with rotating anode operated at 40 kV with 75 mA Small angle X-ray scattering (SAXS, Tokyo, Japan) was employed to reveal insights regarding the pore morphology, as well as to calculate the specific surface area. The electrochemical activity imaging was carried out under an ElProScan 3 Series Scanning Electrochemical Microscope (HEKA division of Harvard Bioscience Inc., Lambrecht (Pfalz), Germany) grounded in a Faraday cage and placed on an antivibration table, as well as with a piezoelectric nanopositioning system for fine approaches. The gold samples were soldered into a plastic Petri-dish (60×15 mm) with a back electrical connection suitable for kinetic studies that require substrate polarization. The SECM measurements were performed using a Pt counter electrode, Ag/AgCl reference electrode, and Pt ultramicroelectrode (UME) with a diameter of 10 μm and an RG of 5 (ratio between the insulator thickness (glass capillary that surrounds the electrode) and the radius of the electrode). The redox mediator solution employed in the experiments contained 3 mM FcMeOH (ferrocenemethanol) in 0.1 M KCl dissolved in ultrapure water. The reagents were purchased from Sigma-Aldrich (Darmstadt, Germany) and Alfa Aesar (Kandel, Germany), respectively. All chemicals were used without further purification.

3. Results and Discussions

3.1. Micro-Structural and Optical Investigations

The photographs presented as insets in Figure 1 show that the gold color changes with the increase in etching time, attesting to the shift of the metallic behavior towards longer wavelengths because of the decrease in the apparent plasma frequency caused by the shorter relaxation time of free carriers related to the scattering process in the material [30].

Thus, it can be clearly seen that, if the as deposited gold layer was initially reflective, after 1000 s of anodization, it becomes a perfect absorber in the visible spectrum. The SEM micrographs offer a thorough view of the gold anodization process evolution. Accordingly, the white islands present in the top-view images of the samples NPG200 and NPG400 are gold clusters formed on the sample surfaces by the removed atoms from the bottom of the pores. Advancing with the etching time, even though the color of the samples fabricated at 600, 800, and 1000 s only slightly increases in darkness, the analysis of the SEM micrographs using ImageJ software (v1.52a, NIH, USA) [31] reveals that the gold solid surface area decreases from 67% for NPG600 to 44% for NPG1000 (Figure S2). Moreover, the threshold level used in ImageJ for estimating the fractal dimension with the box counting method of NPG is depicted in Figure S2.

The gold anodization process recording correlated with the corresponding SEM images is also provided (Video S1). The in-depth evolution of the etching process is displayed in the transversal-view SEM images, where it can be observed that the thickness of the resulting porous layers increases from approximately 100 nm @ 200 s to 230 nm @ 1000 s, while the average pore surface area increased from almost 3 to 43% (Figure 1). As a result, the change of the color appearance with etching time is determined both by the increased porosity and the increase of the porous layer thickness.

Figure 2a shows typical voltage–time curves at a constant current density for the anodization of the Au substrate. Initially, voltage jumps (approximately 2.5 V) were observed corresponding to passive-to-active transition [32], followed by a temporary stagnation for a short period where an equilibration of the electrochemical system occurs. Thereafter, the voltage profile has a periodical saw-like shape profile that corresponds to a typical two processes' competition. The voltage rise is associated with the formation of barrier oxide, whereas the voltage decay corresponds to the localized etching of the barrier layer that ultimately leads to the gold layer porosification [33]. Accordingly, the spikes in the voltage indicate practically when corrugations' boundaries are "attacked" by HF, thus triggering the porosification process. As the process time increases, an increasing tendency is observed up to 400 s, where initialization of different porosification sites takes place, followed by a plateau characterized by in-depth pores' evolutions, where an intense gas evolution from the sample surface was observed. The formation of pores is governed by the diffusion of reagents and is limited by the diffusion, as the agitation was not used during the etching process. A schematic illustration of this process mechanism is presented in Figure 2b. Thus, under anodic bias, an oxide layer is formed on the gold surface by consuming water to form $Au(OH)_x$ and H^+ ions [3]—Figure 2b(i). Because of the surface corrugations, the electric field is stronger on top than at the bottom (at corrugations' boundaries) (Figure 2b(ii)), increasing the reaction rate on top, thus leading to a thicker oxide layer on those areas

(Figure 2b(iii)). The second reaction is represented by dissolving the oxide layer by the HF, this time with the reaction being faster at the bottom of corrugation because of the thinner oxide layer and the increased concentration of H^+ transported from the top by diffusion and convection (Figure 2b(iv)). Whereas, during silicon porosification, a fluorosilane layer typically forms, AuF_4 or AuF_3 are formed in this case [34]. These compounds are instable and can decompose easily as $Au(OH)_3$ in aqueous basic solution [35]. In order to increase the selective removal of $Au(OH)_x$ at the etching front, DMF was used as electrolyte to increase the stability of two compounds, AuF_4 and AuF_3 [36]. Therefore, at the etching front, a balance was established between oxidation and dissolution in that matter, that the resulting effect is the etching into depth of the gold surface, while the pore walls are protected by the reforming oxide (Figure 2b(v)). The difference between the two reactions' speeds on top of the pore and at the bottom lead to the pore growing into depth (Figure 2b(vi)). The in-depth porosification practically represents a cyclic alternance of the (iii–vi) reaction steps. After the reaction stops, the remaining oxide layer is dissolved by the HF.

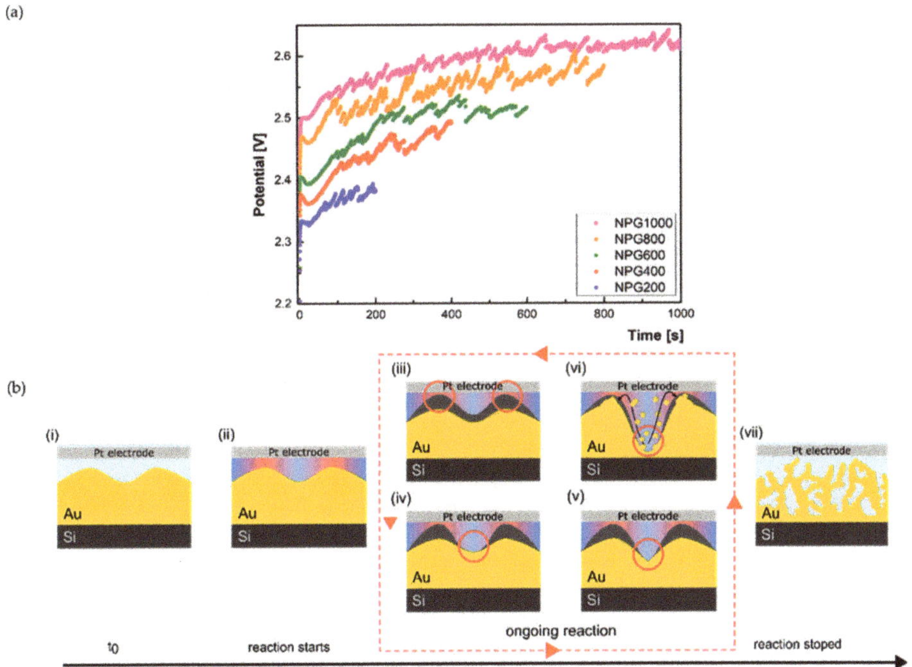

Figure 2. (**a**) The potential variation with etching time for a constant intensity of 15 mA (for clarity, the experimental points were translated vertically by 0.05 V) and (**b**) nanoporous gold formation reaction (i–vii).

It is worth mentioning that we made a step further in validation of the proposed technology for the fabrication of the nanoporous gold, and we verified the robustness of the method for large-scale practical applications. Thus, using a 4 inch Si wafer size porosification system, we demonstrated the reproducibility of the proposed process, obtaining a highly uniform NPG layer on the roughly entire surface (Figure S3) that can then be subjected to standard photolithographic processes for more complex device fabrication.

The optical properties of the nanoporous gold samples were investigated measuring the absorption spectra (Figure 3).

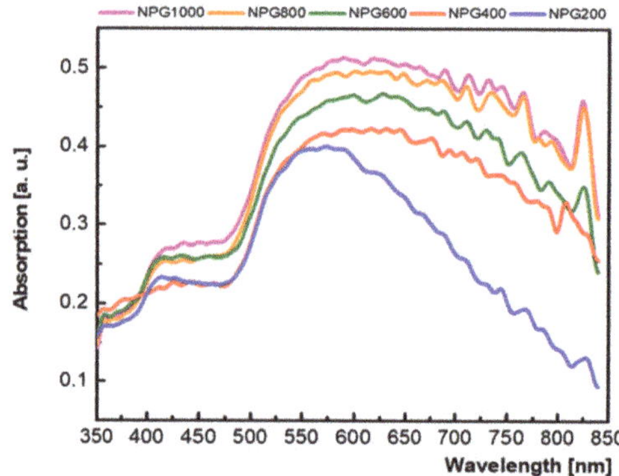

Figure 3. Absorption spectra of the NPG samples (the NPG spectra are normalized to the flat gold film absorption spectrum).

Two peaks are clearly defined, the first one in the 400–450 nm region that corresponds to the bulk gold plasmonic peak, and the second one at a wavelength beyond 500 nm assigned to localized surface plasmons, which depends on the morphology of the structure [37]. Taking into account that the NPG structures present a randomly distributed and different sizes of gold ligaments that support additional plasmonic modes, the broad absorption peak from longer wavelengths suffers a red shift with the increase of etching time. Moreover, the plasmonic resonances become stronger and the absorption peak increases. This effect could be attributed to the increase of both the porosity and the porous layer thickness (i.e., a lower porosity implies larger gold diameters, which, similar to the case of nanoparticles and nanorods, leads to a shift of the resonance wavelength).

Further, to complete the SEM surface investigations and to unravel the volumetric characteristics in terms of porosity and surface area of NPG, the SAXS analysis was employed.

Thus, for acquisition of the SAXS patterns, the incidence angle, θ, was varied from 0 to 2°. The scattering vector, q, defined using scattering angle (θ) and the incident X-ray wavelength ($\lambda = 1.5406$ Å), was calculated using Equation (1):

$$q = \frac{4\pi sin\theta}{\lambda} \tag{1}$$

The resulting SAXS patterns of the intensity scattered by the investigated samples, expressed as a function of the scattering vector q, are presented in Figure 4.

Generally, two distinct domains arise in the SAXS patterns: on the one hand, the low-q scattering domain encodes the characteristics of crystalline domains examining the radius of gyration (R_G) and, on the other hand, the high-q domain obeys a power law and is related to the fractal dimension of the samples [38]. For instance, the slope (s) of high-q region ranges between −3.66 and −3.72, leading to a decrease of the fractal dimension, D, calculated as 6-s, from 2.34 to 2.27, which is related to an increase in the nanostructuration level.

In addition, it can also be observed that each SAXS pattern exhibits critical scattering vectors, where the intensity drops, which can be ascribed to the sample porosity. The position of the critical angle corresponding to bulk gold (θ_{c-Au}) is preserved (illustrated with blue line), while the one related to the porous gold layer (θ_{c-NPG}) suffers a shift towards smaller q values. The position of the critical scattering vector in the reciprocal space is: $q_c = 0.092$ Å^{-1} (blue dashed line in Figure 4 corresponds to bulk gold density, $\rho_{Au} = 19.3$ g/cm$^{3)}$. In the following, this critical angle will be denoted as q_{c-Au}. At the same time, the values of the critical scattering vector that correspond to nanoporous gold q_{c-NPG} successively decrease from 0.074 Å^{-1} (NPG200), 0.065 Å^{-1} (NPG400), 0.064 Å^{-1} (NPG600), 0.062 Å^{-1}

(NPG800), and up to 0.055 Å^{-1} for NPG1000. As the critical scattering vector for porous gold does not have a fixed value, the above values can be considered as average ones with a deviation of ±0.002 Å^{-1}.

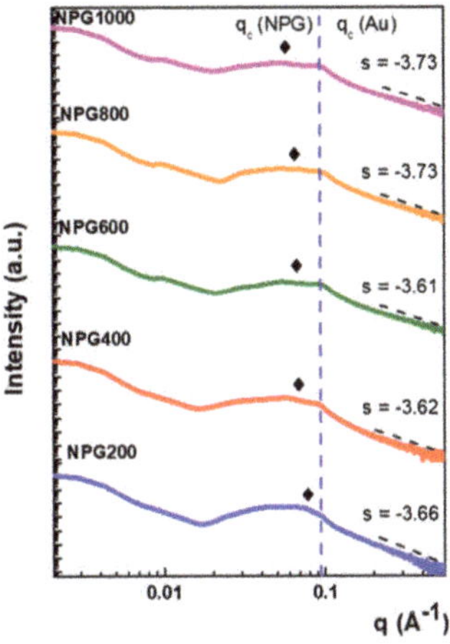

Figure 4. Small angle X-ray scattering (SAXS) patterns recorded for the investigated NPG samples. The blue dashed line indicates the critical angle of the gold, while diamonds stand for the corresponding ones for porous gold.

To relate the critical scattering vector to the sample density, the following calculations were performed considering that the complex index of refraction of matter for X-rays is close to unity [39]:

$$n = 1 - \delta - i\beta \tag{2}$$

where δ is determined by the dispersion and β is proportional with the absorption coefficient having the following values [40]:

$$\delta = \frac{\lambda^2}{2\pi} r_e \rho \tag{3}$$

$$\beta = \frac{\lambda}{4\pi} \mu, \tag{4}$$

where r_e is the classical electron radius ($r_e = 2.8 \times 10^{-15}$ m), ρ is the electron density, and μ is the linear absorption coefficient. In the assumption that the investigated materials has no absorption ($\beta = 0$), the refractive index is expressed as follows:

$$n_{layer2} = 1 - \delta \tag{5}$$

and the angle of the total reflection in the real space is given by the following [40,41]:

$$\theta_c = \sqrt{2\delta} \tag{6}$$

Taking into account the relationship between the critical angle and the critical scattering vector (Equation (1)), as well as Equation (3), which relates δ coefficient by the sample density, we can express the sample density in the reciprocal space as follows:

$$\rho = \frac{\left(asin\left(\frac{\lambda q_c}{4\pi}\right)\right)^2}{2} \frac{2\pi}{\lambda^2 r_e} \tag{7}$$

Using the values obtained for the critical scattering vectors as well as its variation, the density of nanoporous gold was estimated, showing a decrease from 12.5 ± 0.2 g/cm^3 to 6.29 ± 0.1 g/cm^3. Finally, the NPG layer's porosity was determined as follows:

$$p = 1 - \frac{\rho_{NPG}}{\rho_{Au}} \tag{8}$$

Table 1 summarizes the values obtained for density and porosity calculated from SAXS patterns, as well as for average solid surface area ($\overline{p_{SEM}}$) and 2D fractal dimension (D_{SEM}) obtained from analysing the SEM micrographs.

Table 1. Morpho-structural properties determined using scanning electronic microscopy (SEM) and X-ray diffraction (XRD) investigations, where ρ_{NPG} is the nanoporous gold density. p and D represent the porosity and fractal dimension of the sample calculated from SAXS patterns, respectively, whereas $\overline{p_{SEM}}$ and D_{SEM} are the average solid surface area and 2D fractal dimension, respectively, obtained from SEM micrographs analysis. NPG, nanoporous gold.

Sample	ρ_{NPG} (g/cm^3)	p (%)	$\overline{p_{SEM}}$ (%)	D	D_{SEM}
NPG200	12.5 ± 0.2	35.2 ± 1	26 ± 3.15	2.34	1.92
NPG400	9.67 ± 0.2	49.9 ± 1	46 ± 1.08	2.38	1.86
NPG600	9.37 ± 0.1	51.5 ± 0.05	56 ± 1.32	2.39	1.76
NPG800	8.79 ± 0.1	54.5 ± 0.05	63 ± 1.6	2.27	1.71
NPG1000	6.29 ± 0.1	67.4 ± 0.05	66 ± 1.81	2.27	1.69

As can be observed, both the porosity and average solid surface area values follow the same trend; the same tendency also arises for the fractal dimensions when determined from SEM and SAXS analyses. The differences between values arise from the probed area in each of the measurements: SEM only locally probes the surface sample morphology, whereas SAXS examines the entire volume of the sample, providing a more realistic view. The smaller values obtained for the fractal dimension by SEM practically confirm this observation, as the surface is generally more affected during the porosification process [42].

Further, to gain a quantitative framework of the specific surface area (S_n), the Porod formalism was used, which is based on the following formulas:

$$Q = \int_0^{\infty} I(q) q^2 dq \tag{9}$$

$$K_p = \lim_{q \to \infty} I(q) q^4 \tag{10}$$

$$S_n = 10^4 \frac{\pi p (1-p)}{\rho} \frac{K_p}{Q} \tag{11}$$

where $I(q)$ is the scattering intensity; Q is the Porod invariant given by the Porod integral (9); K_p is the Porod constant calculated with Formula (10) from the asymptotic behaviour of the tails in the high q region; and p and ρ are the sample porosity and density, respectively.

The Iq versus q and $ln(Iq^4)$ versus q^2 dependences, necessary for the calculation of the Porod integral and Porod constant, respectively, are presented in Figure S4. The results obtained for the Porod integral, Porod constant, and specific surface area are summarized in Table 2.

Table 2. Radius of gyration of ligaments (R_g), size of the crystalline domains ($d_{cryst.}$), Porod integral (Q), Porod constant (K_p), and specific surface area at different times (S_n).

Sample	R_g (nm)	$d_{cryst.}$ (nm)	Q (Å^{-3})	K_p (Å^{-2})	S_n (m^2/g)
NPG200	20.3	57.4	122.75	0.87	3.67 ± 0.18
NPG400	22.1	62.5	59.83	0.61	9.44 ± 0.18
NPG600	18.0	50.9	68.37	1.15	15.77 ± 0.14
NPG800	16.9	47.8	63.27	1.12	17.70 ± 0.12
NPG1000	16.2	45.8	67.81	1.20	17.96 ± 0.11

They show distinct stages in the nanoporous gold layer formation at different etching times. In fact, an increase in the etching time led to significant modifications in the sample porosity and specific surface area (Figure 5). More precisely, with the increasing etching time, the NPG average porosity increases from 35.2 to 67.4%, which is associated with an increase of the specific surface area from 3.3 (for flat gold) to 3.67 for the thinnest NPG200 layer and further up to 17.96 m^2/g for the thickest NPG1000 layer. It can also be observed that the specific surface area for NPG samples obtained at a higher etching time duration presents smaller deviations. This is reasonable taking into account the smaller variations of the porosity in the case of these samples, as shown in Table 1.

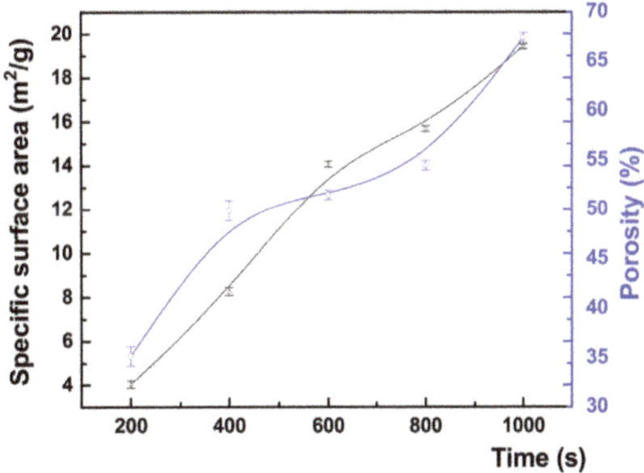

Figure 5. The dependence of the specific surface area (**black line**) and sample porosity (**blue line**) with increasing etching time. Error bars are used to show the standard deviation of the specific surface area and porosity estimated values.

3.2. Electrochemical Characterization

Considering the huge potential of the nanoporous gold films towards electrochemical systems, cyclic voltammetry (CV) and electrochemical impedance spectroscopy (EIS) measurements were performed to evaluate the active surface area and interfacial properties. Thus, the electrochemically addressable surface area was firstly determined using the Au oxide reduction peak measured by CV and the corresponding charge associated with the reduction of gold oxide, by peak integration [43] (five cycles of CV were employed to certify that the response is stabilized).

The cyclic voltammograms were recorded in 0.5 M H_2SO_4 electrolyte, in the potential region from −0.4 to 1.6 V, at the scan rate of 100 mV/s (Figure 6a), and during potential cycling, anodic peaks between 1 and 1.3 V (vs. Ag/AgCl) and a well-defined cathodic peak at ~0.75 V were obtained because of the oxidation and reduction of the outermost layer of gold atoms. Whereas the flat gold surface exhibits a single oxidation peak (Figure 6a) due to the polycrystalline nature of the surface, three partially overlapped peaks arise after porosification determined by the oxide formation on the different crystallographic planes exposed, principally (100), (110), and (111) [44]. The corresponding electrical charge can be obtained by integrating the gold oxide reduction peak, which was converted into the electrochemical surface area (ECSA), assuming 390 μC/cm^2 as the specific charge required for gold oxide reduction for polycrystalline gold [24]. Further, the roughness factor (RF) was calculated normalizing ECSA to the geometrical area (0.5 cm^2). As can be observed in Figure 6b, the effective area of electrodes significantly increases with porosification duration, reaching 15.4 cm^2 for the 1000 s period, revealing an approximately 16-fold enhancement of ECSA and RF compared with the standard flat Au electrode, significantly higher values than those reported for the nanoporous gold obtained via dealloying of Ag-Au alloy layers [45].

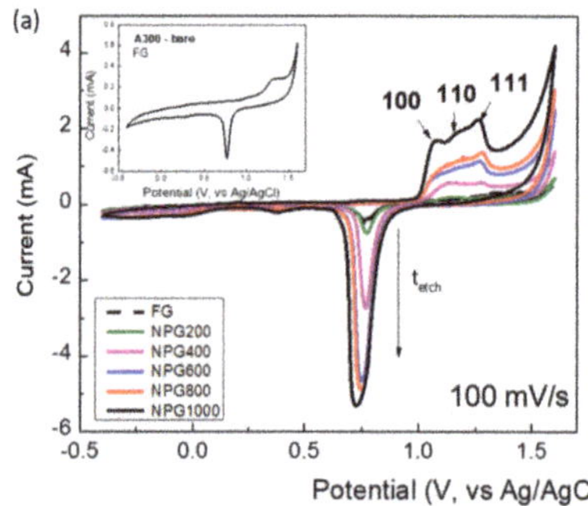
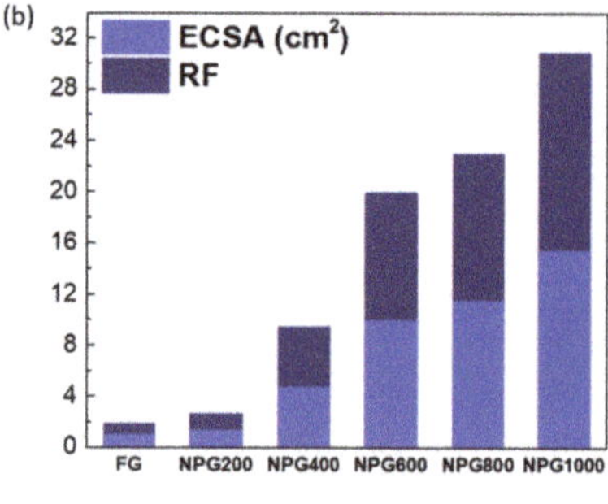

Figure 6. (**a**) Cyclic voltammograms (CVs) recorded with nanoporous Au electrodes in 0.5 M H$_2$SO$_4$ solution, at a scan rate of 100 mV/s (third cycle); (**b**) the electrochemical surface area (ECSA) and the roughness factor (RF) estimated from CVs.

Next, the non-faradaic EIS was used to probe the porous films, measuring the double layer capacitance of the experimental test electrodes. As can be seen in Figure 7a, the Nyquist curves become steeper with the increase of the porosification duration, confirming the increase of capacitance. Concomitantly, the phase maximum successively increases, approaching 80°, and shifts towards lower frequencies—Figure 7b. The capacitive behavior is clearly shown in the intermediary frequency range, where the imaginary part of the impedance presents a linear variation as a function of frequency in log-log scale—Figure 7c. The slope of this linear part corresponds (with opposite sign) to the constant phase element (CPE) exponent that accounts for non-ideal porous electrode capacitance.

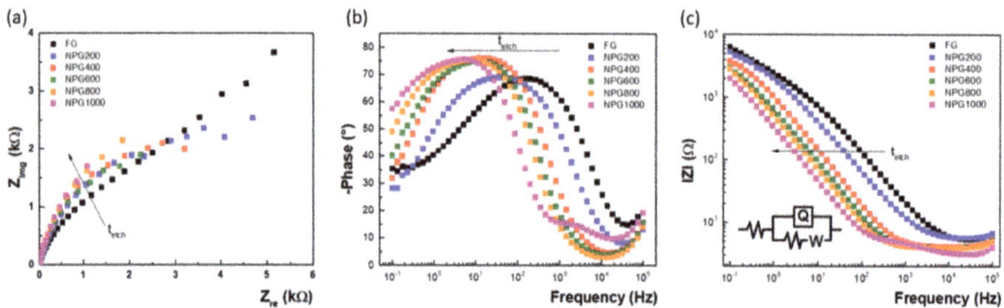

Figure 7. Electrochemical impedance spectroscopy results recorded with the flat Au and nanoporous Au electrodes in 0.5 M H$_2$SO$_4$ solution, at open circuit potential, from 100 kHz to 100 mHz using a 10 mV (rms) sine wave excitation: (**a**) Nyquist plots; (**b**) Bode plots—frequency dependence of the phase; and (**c**) Bode plots—frequency dependence of the imaginary impedance.

The impedance data were fitted considering an equivalent circuit for our system (inset image within Figure 7c containing the ohmic resistance of electrolyte (visible in the high frequency region) and the ionic charge transport resistance through the double layer and the double layer capacitance of the electrode, as well as the restricted diffusion of the electrolyte ions (W) [46] Data analysis was carried out using ZSimpwin software (v3.21, AMETEK, USA). The double layer capacitance values were obtained using the following formula:

$$C_{dl} = Q^{1/n} \cdot R^{(1-n)/n} \tag{12}$$

where Q and n characterize the constant phase element and R is the associated resistance.

Relating the double layer capacitances determined by EIS measurements for the nanoporous gold to that corresponding to standard flat electrode, a roughness factor was also estimated [26], and the values obtained for the investigated electrodes using the two electrochemical techniques are shown in Figure 8. As can be observed, slightly higher values were obtained when the roughness factor was calculated using the EIS data, although they are more appropriate in comparison with previously reported results, where the presence of residual Ag fraction artificially increases the porous film double layer capacitance as its specific capacitance is larger than that of metallic Au [26,27]. As this issue is not present in our case, working with pure gold films, the EIS measurements provide an improved accuracy of values because the excitation signal reaches the bottom of the pores and senses the entire internal area of the porous electrode, not only the sites able to employ oxidation. It is, however, important to note that, strictly comparing our results with similar thickness samples obtained by dealloying processes based on chemical or electrochemical methods [19,47], a substantial simplification of the fabrication protocol was firstly demonstrated, as well as, furthermore, an increased porosity and consequently larger area of the electroactive surface with more than 30%.

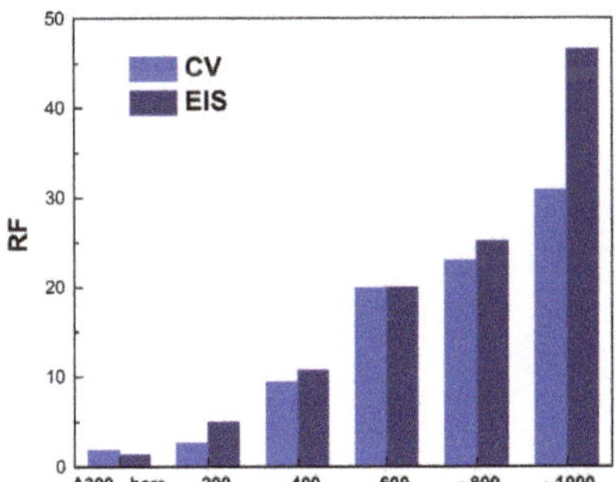

Figure 8. Roughness factor values estimated from CV and electrochemical impedance spectroscopy (EIS) measurements.

To complete the image of nanoporous gold electrodes, faradaic EIS measurements were also performed using ferri-ferrocyanide couple as redox probe aiming to characterize mainly the mass-transport to the electrodes associated with Faradaic processes that occur during charge injection with electrical stimulation. Thus, the EIS was carried out in biased voltage (formal redox potential of the ferri/ferrocyanide redox couple determined from CV measurements) and the recorded data are shown in Figure 9 in both Nyquist and Bode representations.

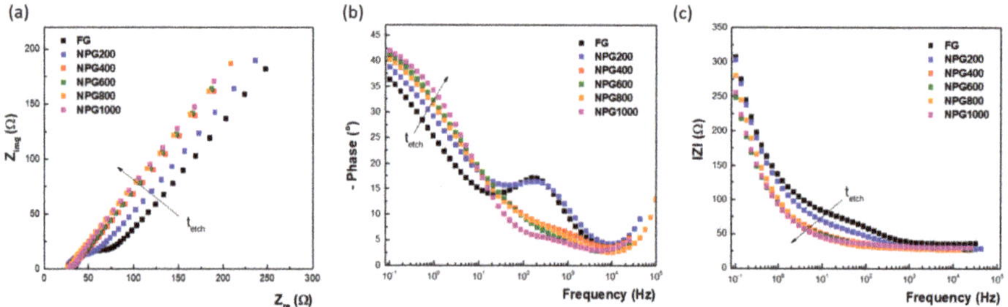

Figure 9. Electrochemical impedance spectroscopy results recorded with the bare Au and nanoporous Au electrodes in 2 mM Fe(CN)$_6$$^{3-}$/Fe(CN)$_6$$^{4-}$ phosphate buffer containing 0.1 M KCl, at the formal potential of the redox probe, from 100 kHz to 100 mHz, using a 10 mV (rms) sine wave excitation: (**a**) Nyquist plots; (**b**) Bode plots—frequency dependence of the phase; and (**c**) Bode plots—frequency dependence of the impedance module.

Remarkable differences can be observed when the porosification process is applied to obtain different thicknesses of the nanoporous gold layer. Thus, if the Nyquist plot initially consists of a distinct semicircle counting for the charge transfer resistance, which corresponds to the Bode phase peak in the intermediary frequency range, the radius of the semicircle starts to diminish after 200 s of porosification, moving the phase peak towards higher frequencies, and it practically disappears at higher process durations. In this case, we can talk about an extremely low electrode impedance, determined by the high surface area, and a correspondingly large charge injection capacity. Thus, the gold electrode nanostructuration led to enhanced surface area and conductivity.

The fitting of experimental data was done using a similar equivalent circuit, where the main parameter in the case of the faradaic EIS is the electron transfer resistance, which controls the electron transfer kinetics of the redox probe at the electrode interface. As a result, the heterogeneous electron transfer rate constant across the interface was obtained through the following relation [48]:

$$k^0 = \frac{RT}{n^2 F^2 A R_{ct} c}$$

(13)

where R is the gas constant, T is temperature in Kelvin, n is the number of moles of electrons transferred in the redox reaction ($n = 1$ for Fe(CN)$_6$$^{3-/4-}$), F is the Faraday constant, A is the geometrical area of the electrode (0.5 cm^2), R_{ct} is the charge transfer resistance determined from EIS, and c is the concentration of the redox couple.

The obtained charge transfer resistance suggests the easing of charge transfer to and from the porous electrodes, also reflected by the continuous increase of k^0 values, which are larger in comparison with the previously reported results [48]. Thus, the heterogeneous electron transfer rate constant becomes 50 times higher for the thickest gold nanoporous layer, a similar order of increase as for the roughness factor (Figure 10).

Finally, the powerful spatial resolution of the scanning electrochemical microscopy technique [49] was employed as a cross-checking kinetic tool in the study of the nanoporous gold electrodes, practically achieving a correlation between the surface morphology investigated by SEM and the electrochemical activity. Experiments were performed using a 10 µm diameter Pt disk ultramicroelectrode in a 1 mM FcMeOH (0.1 M KCl electrolyte) solution.

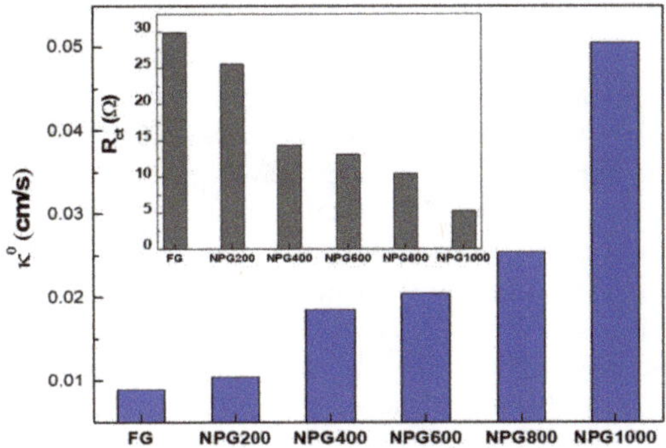

Figure 10. Heterogeneous electron transfer rate constant and electron transfer resistance (inset graph) determined from faradaic EIS measurements.

As shown in Figure 11, electrochemical details are unveiled in a spatially-resolved manner with the emphasis on the differences between the two substrates electrochemical response. Here, the 3D scans were performed over the nanoporous gold obtained after a 800 s anodization process in comparison with the initial flat substrate, showing the differences in electrochemical response in the presence of the redox mediator of the two unbiased substrates. It is worth mentioning here that the porous counterpart has higher current feedback zones because of the fact that many of the features formed during the process are suitable for the reduction of [FcMeOH]$^+$. Moreover, the ligaments and the micropores enable a large surface area that drastically inhibits mass transfer of educts and products [50], which translates into a drop of feedback current as the tip scans over the surface. At the same time, the slow mass transfer on the overall reactivity of the ligaments and micropores makes the nanoporous gold substrate very attractive to many electrochemical applications. Therefore, the electrochemical investigations demonstrate the huge potential of nanoporous gold electrodes to be used for both non-faradaic electrochemical studies, where, generally, 1 kHz low impedance is required for neural electrophysiology to reduce the noise levels [51], and high sensitivity label-free affinity binding in faradaic-based biosensing, where high heterogeneous electron transfer rate constant is required [52].

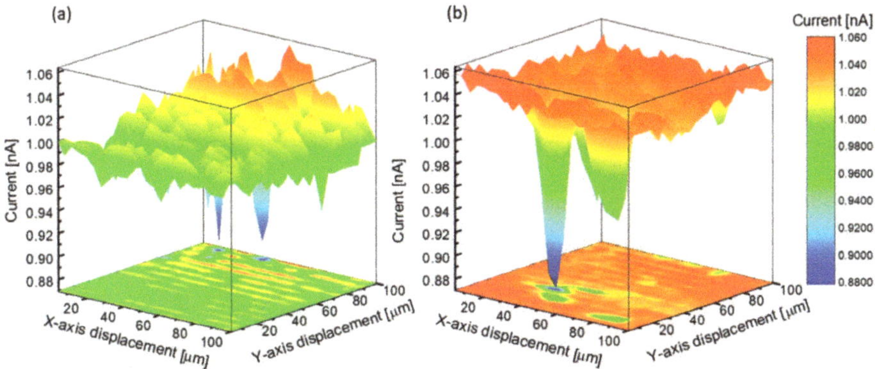

Figure 11. Scanning electrochemical microscopy (SECM) mapping images (100 × 100 µm) obtained in feedback mode when a tip potential of 0.35 V was applied: (**a**) Flat Gold (**b**) NPG800.

4. Conclusions

We report a cost-effective, easily scalable, and controllable method to fabricate nanoporous gold films directly on silicon substrate for optical and/or electrochemical sensor applications that can be applied to other types of substrates, such as glass or flexible polymeric substrates. Moreover, adjusting the anodization current density, the nanoporous gold ligaments would be fine-tuned in order to achieve the desired properties. The careful calibration of the NPG fabrication process was complemented by a systematic characterization of the experimental samples in order to assess the evolution of the morpho-structural, optical, and electrochemical properties. The main drawbacks that other methods have, such as higher cost of fabrication, additional procedures of transferring onto a substrate, and residual material that could alter its properties, were overcome by this method of anodization at a lower voltage. The investigations show on the one hand the progressive increase of the sample porosity and specific surface area along with the etching time, and on the other hand the achievement of tunable optical and electrochemical performances that open new opportunities for the consistent development of novel devices based on NPG using a reliable wafer-scale process.

Supplementary Materials: The following are available online at http://www.mdpi.com/2079-4991/10/11/2321/s1, Figure S1: Experimental setup for nanoporous gold fabrication; schematic representation of the NPG etching system–cross-sectional view, Figure S2: The calculated average gold solid area (from SEM micrographs) for different etching time durations and the binarized images (insets), which were used for calculation of the fractal dimension in ImageJ by using box counting method, Figure S3: (a) 4 inch NPG on Si fabrication using AMMT system; (b) Photography of the resulted on wafer NPG film in comparison with the one obtained using microcell. Insets represents SEM top views of NPG in two distinct regions of the wafer, Figure S4: Iq vs. q and ln(Iq4) vs. q2 dependences for the: (a) Porod integral and (b) Porod constant, Video S1: Fabrication of NPG.

Author Contributions: Conceptualization, P.V. and M.K.; Investigation, C.R., A.B., O.T., G.C., I.M., and M.K.; Methodology, A.R.; Writing—original draft, P.V.; Writing—review & editing, P.V. and M.K. All authors have read and agreed to the published version of the manuscript.

Funding: Financial support was offered by the Core Program PN 1916/2019 MICRO-NANO-SIS PLUS/08.02.2019 and PN-III-P1-1.2-PCCDI-2017-0820-Project no. 1, funded by MCI. Mihalache's work was supported by a grant of the Romanian Ministry of Education and Research, CNCS-UEFISCDI, project number PN-III-P1-1.1-PD-2019-1081, within PNCDI III.

Conflicts of Interest: The authors declare no conflict of interest.

References

1. Forty, A.J. Corrosion micromorphology of noble metal alloys and depletion gilding. *Nature* **1979**, *282*, 597–598. [CrossRef]
2. Pickering, H.W.; Swann, P.R. Electron Metallography of Chemical Attack Upon Some Alloys Susceptible to Stress Corrosion Cracking. *Corrosion* **1963**, *19*, 373t–389t. [CrossRef]
3. Fang, C.; Bandaru, N.M.; Ellis, A.V.; Voelcker, N.H. Electrochemical fabrication of nanoporous gold. *J. Mater. Chem.* **2012**, *22*, 2952–2957. [CrossRef]
4. Wittstock, A.; Biener, J.; Bäumer, M. Nanoporous gold: A new material for catalytic and sensor applications. *Phys. Chem. Chem. Phys.* **2010**, *12*, 12919–12930. [CrossRef]
5. Xiao-ting, Z.H.U.; Lu-jia, Z.; Hong, T.A.O.; Jun-wei, D.I. Synthesis of Nanoporous Gold Electrode and Its Application in Electrochemical Sensor. *Chin. J. Anal. Chem.* **2013**, *41*, 693–697. [CrossRef]
6. Collinson, M.M. Nanoporous Gold Electrodes and Their Applications in Analytical Chemistry. *Int. Sch. Res. Not.* **2013**, *2013*, 1–21. [CrossRef]
7. Lee, K.U.; Byun, J.Y.; Shin, H.J.; Kim, S.H. A high-performance supercapacitor based on polyaniline-nanoporous gold. *J. Alloys Compd.* **2019**, *779*, 74–80. [CrossRef]
8. Ferris, A.; Bourrier, D.; Garbarino, S.; Guay, D.; Pech, D. 3D Interdigitated Microsupercapacitors with Record Areal Cell Capacitance. *Small* **2019**, *15*, 1901224. [CrossRef] [PubMed]
9. Kim, S.I.; Kim, S.W.; Jung, K.; Kim, J.B.; Jang, J.H. Ideal nanoporous gold based supercapacitors with theoretical capacitance and high energy/power density. *Nano Energy* **2016**, *24*, 17–24. [CrossRef]
10. Garoli, D.; Calandrini, E.; Giovannini, G.; Hubarevich, A.; Caligiuri, V.; De Angelis, F. Nanoporous gold metamaterials for high sensitivity plasmonic sensing. *Nanoscale Horizons* **2019**, *4*, 1153–1157. [CrossRef]

11. Garoli, D.; Calandrini, E.; Bozzola, A.; Toma, A.; Cattarin, S.; Ortolani, M.; De Angelis, F.; Fisica, D.; Universita, S.; Moro, P.A. Fractal-Like Plasmonic Metamaterial with a Tailorable Plasma Frequency in the near-Infrared. *ACS Photonics* **2018**, *5*, 3408–3414. [CrossRef]

12. Falsig, H.; Hvolbæk, B.; Kristensen, I.S.; Jiang, T.; Bligaard, T.; Christensen, C.H.; Nørskov, J.K. Trends in the catalytic CO oxidation activity of nanoparticles. *Angew. Chem. Int. Ed.* **2008**, *47*, 4835–4839. [CrossRef] [PubMed]

13. Xu, C.; Su, J.; Xu, X.; Liu, P.; Zhao, H.; Tian, F.; Ding, Y. Low temperature CO oxidation over unsupported nanoporous gold. *J. Am. Chem. Soc.* **2007**, *129*, 42–43. [CrossRef] [PubMed]

14. Zielasek, V.; Jürgens, B.; Schulz, C.; Biener, J.; Biener, M.M.; Hamza, A.V.; Bäumer, M. Gold catalysts: Nanoporous gold foams. *Angew. Chem. Int. Ed.* **2006**, *45*, 8241–8244. [CrossRef]

15. Qadir, K.; Quynh, B.T.P.; Lee, H.; Moon, S.Y.; Kim, S.H.; Park, J.Y. Tailoring metal–oxide interfaces of inverse catalysts of TiO 2 /nanoporous-Au under hydrogen oxidation. *Chem. Commun.* **2015**, *51*, 9620–9623. [CrossRef] [PubMed]

16. Bond, G.C.; Village, N.T.; Hill, W.; Rg, R. Gold-Catalysed Oxidation of and David T Thompson. *Gold Bull.* **2000**, *33*, 41–50. [CrossRef]

17. Kosinova, A.; Wang, D.; Baradács, E.; Parditka, B.; Kups, T.; Klinger, L.; Erdélyi, Z.; Schaaf, P.; Rabkin, E. Tuning the nanoscale morphology and optical properties of porous gold nanoparticles by surface passivation and annealing. *Acta Mater.* **2017**, *127*, 108–116. [CrossRef]

18. Ruffato, G.; Garoli, D.; Cattarin, S.; Barison, S.; Natali, M.; Canton, P.; Benedetti, A. Microporous and Mesoporous Materials Patterned nanoporous-gold thin layers: Structure control and tailoring of plasmonic properties. *Microporous Mesoporous Mater.* **2012**, *163*, 153–159. [CrossRef]

19. Kamiuchi, N.; Sun, K.; Aso, R.; Tane, M.; Tamaoka, T.; Yoshida, H.; Takeda, S. Self-Activated surface dynamics in gold catalysts under reaction environments. *Nat. Commun.* **2018**, *9*, 1–6. [CrossRef]

20. Calandrini, E.; Giovannini, G.; Garoli, D. 3D nanoporous antennas as a platform for high sensitivity IR plasmonic sensing. *Opt. Express* **2019**, *27*, 25912. [CrossRef]

21. Jia, F.; Yu, C.; Ai, Z.; Zhang, L. Fabrication of Nanoporous Gold Film Electrodes with Ultrahigh Surface Area and Electrochemical Activity. *Chem. Mater.* **2007**, *19*, 3648–3653. [CrossRef]

22. Rebbecchi, T.A.; Chen, Y. Template-based fabrication of nanoporous metals. *J. Mater. Res.* **2018**, *33*, 2–15. [CrossRef]

23. Patel, J.; Radhakrishnan, L.; Zhao, B.; Uppalapati, B.; Daniels, R.C.; Ward, K.R.; Collinson, M.M. Electrochemical Properties of Nanostructured Porous Gold Electrodes in Biofouling Solutions. *Anal. Chem.* **2013**, *85*, 11610–11618. [CrossRef] [PubMed]

24. Liu, L.; Lee, W.; Huang, Z.; Scholz, R. Fabrication and characterization of a flow-through nanoporous gold nanowire / AAO composite membrane. *Nanotechnology* **2008**, *19*, 335604. [CrossRef]

25. Beets, N.; Stuckner, J.; Murayama, M.; Farkas, D. Fracture in nanoporous gold: An integrated computational and experimental study. *Acta Mater.* **2020**, *185*, 257–270. [CrossRef]

26. Mattarozzi, L.; Cattarin, S.; Comisso, N.; Guerriero, P.; Musiani, M.; Verlato, E. Electrochimica Acta Preparation of porous nanostructured Ag electrodes for sensitive electrochemical detection of hydrogen peroxide. *Electrochim. Acta* **2016**, *198*, 296–303. [CrossRef]

27. Rouya, E.; Cattarin, S.; Reed, M.L.; Kelly, R.G.; Zangari, G.; Energetica, L.; Uniti, C.S. Electrochemical Characterization of the Surface Area of Nanoporous Gold Films. *J. Electrochem.* **2012**, *159*, K97. [CrossRef]

28. Nishio, K.; Masuda, H. Anodization of Gold in Oxalate Solution To Form a Nanoporous Black film. *Angew. Chem. Int. Ed. Engl.* **2011**, *50*, 1603–1607. [CrossRef]

29. Simion, M.; Kleps, I.; Neghina, T.; Angelescu, A.; Miu, M.; Bragaru, A.; Danila, M.; Condac, E.; Costache, M.; Savu, L. Nanoporous silicon matrix used as biomaterial. *J. Alloys Compd.* **2007**, *434–435*, 830–832. [CrossRef]

30. Ruffato, G.; Romanato, F.; Garoli, D.; Cattarin, S. Nanoporous gold plasmonic structures for sensing applications. *Optics Express* **2011**, *19*, 13164–13170. [CrossRef]

31. Schneider, C.A.; Rasband, W.S.; Eliceiri, K.W. HISTORICAL commentary NIH Image to ImageJ: 25 years of image analysis. *Nat. Methods* **2012**, *9*, 671–675. [CrossRef] [PubMed]

32. Asoh, H.; Uehara, A.; Ono, S. Fabrication of porous silica with three-dimensional periodicity by Si anodization. *Jpn. J. Appl. Phys. Part 2 Lett.* **2004**, *43*, 43–46. [CrossRef]

33. Parkhutik, V.; Curiel-Esparza, J.; Millan, M.C.; Albella, J. Spontaneous layering of porous silicon layers formed at high current densities. *Phys. Status Solidi Appl. Mater. Sci.* **2005**, *202*, 1576–1580. [CrossRef]

34. Zhang, X.G. *Electrochemistry of Silicon and Its Oxide*; Springer Science & Business Media: Heidelberg/Berlin, Germany, 2007.

35. Mohr, F. The chemistry of gold-fluoro compounds: A continuing challenge for gold chemists. *Gold Bull.* **2004**, *37*, 164–169. [CrossRef]

36. Fang, C.; Ellis, A.V.; Voelcker, N.H. Electrochemically prepared porous silver and its application in surface-enhanced Raman scattering. *J. Electroanal. Chem.* **2011**, *659*, 151–160. [CrossRef]

37. Lang, X.; Qian, L.; Guan, P.; Zi, J.; Chen, M. Localized surface plasmon resonance of nanoporous gold. *Appl. Phys. Lett.* **2011**, *98*, 2009–2012. [CrossRef]

38. Romanitan, C.; Varasteanu, P.; Mihalache, I.; Culita, D.; Somacescu, S.; Pascu, R.; Tanasa, E.; Eremia, S.A.V.; Boldeiu, A.; Simion, M.; et al. High-performance solid state supercapacitors assembling graphene interconnected networks in porous silicon electrode by electrochemical methods using 2,6-dihydroxynaphthalen. *Sci. Rep.* **2018**, *8*, 1–14. [CrossRef]

39. Bowen, K.; Tanner, B. *High Resolution X-ray Diffractometer and Topography*; Taylor & Francis: London, UK, 1998; ISBN 0850667585.

40. Daillant, J.; Gibaud, A. *X-Ray and Neutron Reflectivity: Principles and Applications*; Springer: Berlin/Heidelberg, Germany, 1999; ISBN 3540661956.

41. Birkholz, M.; Birkholz, M.; Fewster, P.F.; Genzel, C. *Thin Film Analysis by X-Ray Scattering*; John Wiley & Sons: Hoboken, NJ, USA, 2006; ISBN 9783527310524.

42. Miu, M.; Danila, M.; Kleps, I.; Bragaru, A.; Simion, M. Nanostructure and internal strain distribution in porous silicon. *J. Nanosci. Nanotechnol.* **2011**, *11*, 9136–9142. [CrossRef]

43. Xiao, X.; Si, P.; Magner, E. An overview of dealloyed nanoporous gold in bioelectrochemistry. *Bioelectrochemistry* **2016**, *109*, 117–126. [CrossRef]

44. Jeyabharathi, C.; Ahrens, P.; Hasse, U.; Scholz, F. Identification of low-index crystal planes of polycrystalline gold on the basis of electrochemical oxide layer formation. *J. Solid State Electrochem.* **2016**, *20*, 3025–3031. [CrossRef]

45. Valenti, G.; Scarabino, S.; Goudeau, B.; Lesch, A.; Jović, M.; Villani, E.; Sentic, M.; Rapino, S.; Arbault, S.; Paolucci, F.; et al. Single Cell Electrochemiluminescence Imaging: From the Proof-of-Concept to Disposable Device-Based Analysis. *J. Am. Chem. Soc.* **2017**, *139*, 16830–16837. [CrossRef] [PubMed]

46. Bisquert, J.; Garcia-Belmonte, G.; Bueno, P.; Longo, E.; Bulhões, L.O.S. Impedance of constant phase element (CPE)-blocked diffusion in film electrodes. *J. Electroanal. Chem.* **1998**, *452*, 229–234. [CrossRef]

47. Francis, N.J.; Knospe, C.R. Fabrication and Characterization of Nanoporous Gold Electrodes for Sensor Applications. *Adv. Eng. Mater.* **2019**, *21*, 1–13. [CrossRef]

48. Bae, J.H.; Yu, Y.; Mirkin, M.V. Scanning Electrochemical Microscopy Study of Electron-Transfer Kinetics and Catalysis at Nanoporous Electrodes. *J. Phys. Chem. C* **2016**, *120*, 20651–20658. [CrossRef]

49. Bard, A.J.; Fan, F.R.F.; Kwak, J.; Lev, O. Scanning Electrochemical Microscopy. Introduction and Principles. *Anal. Chem.* **1989**, *61*, 132–138. [CrossRef]

50. Haensch, M.; Balboa, L.; Graf, M.; Silva Olaya, A.R.; Weissmüller, J.; Wittstock, G. Mass Transport in Porous Electrodes Studied by Scanning Electrochemical Microscopy: Example of Nanoporous Gold. *ChemElectroChem* **2019**, *6*, 3160–3166. [CrossRef]

51. Seker, E.; Berdichevsky, Y.; Begley, M.R. The fabrication of low-impedance nanoporous gold multiple-electrode arrays for neural electrophysiology studies. *Nanotechnology* **2010**, *21*, 125504. [CrossRef]

52. Hu, K.; Lan, D.; Li, X.; Zhang, S. Electrochemical DNA Biosensor Based on Nanoporous Gold Electrode and Multifunctional Encoded DNA-Au Bio Bar Codes. *Anal. Chem.* **2008**, *80*, 9124–9130. [CrossRef]

Publisher's Note: MDPI stays neutral with regard to jurisdictional claims in published maps and institutional affiliations.

Article

Effect of Nano Copper on the Densification of Spark Plasma Sintered W–Cu Composites

Vadde Madhur [1,†], Muthe Srikanth [1], A. Raja Annamalai [1,†], A. Muthuchamy [2], Dinesh K. Agrawal [3] and Chun-Ping Jen [4,*]

[1] Centre for Innovative Manufacturing Research, VIT, Vellore, Tamil Nadu 632 014, India; madhur.vadde@gmail.com (V.M.); muthe.srikanth@vit.ac.in (M.S.); raja.annamalai@vit.ac.in (A.R.A.)

[2] Department of Metallurgical and Materials Engineering, National Institute of Technology Tiruchirappalli, Tamil Nadu 620015, India; muthuchamy@nitt.edu

[3] Materials Research Institute, The Pennsylvania State University, University Park, PA 16802, USA; dxa4@psu.edu

[4] Department of Mechanical Engineering and Advanced Institute of Manufacturing for High-Tech Innovations, National Chung Cheng University, Chia-Yi 62102, Taiwan

* Correspondence: imecpj@ccu.edu.tw

† These authors contributed equally to this work.

Abstract: In the present work, nano Cu (0, 5, 10, 15, 20, 25 wt.%) was added to W, and W–Cu composites were fabricated using the spark plasma sintering (S.P.S.) technique. The densification, microstructural evolution, tensile strength, micro-hardness, and electrical conductivity of the W–Cu composite samples were evaluated. It was observed that increasing the copper content resulted in increasing the relative sintered density, with the highest being 82.26% in the W75% + Cu25% composite. The XRD phase analysis indicated that there was no evidence of intermetallic phases. The highest ultimate (tensile) strength, micro-hardness, and electrical conductivity obtained was 415 MPa, 341.44 $HV_{0.1}$, and 28.2% IACS, respectively, for a sample containing 25 wt.% nano-copper. Fractography of the tensile tested samples revealed a mixed-mode of fracture. As anticipated, increasing the nano-copper content in the samples resulted in increased electrical conductivity.

Keywords: tungsten-(*nano*) copper composites; solid-state sintering; spark plasma sintering; microstructure; mechanical properties

Citation: Madhur, V.; Srikanth, M.; Annamalai, A.R.; Muthuchamy, A.; Agrawal, D.K.; Jen, C.-P. Effect of Nano Copper on the Densification of Spark Plasma Sintered W–Cu Composites. *Nanomaterials* **2021**, *11*, 413. https://doi.org/10.3390/nano11020413

Academic Editor: Mirela Suchea

Received: 8 January 2021
Accepted: 1 February 2021
Published: 5 February 2021

Publisher's Note: MDPI stays neutral with regard to jurisdictional claims in published maps and institutional affiliations.

1. Introduction

Refractory metal tungsten (W) has excellent thermal creep resistance, high electrical and thermal conductivity, and the lowest vapor pressure among all metals. On the other hand, copper (Cu) has superior electrical and thermal conductivity and has good corrosion resistance. When copper is added to tungsten, a binary system in which the two metals are mutually insoluble [1] is formed. This insolubility is due to incompatibility in their crystal structures. W is body-centered cubic (BCC), and Cu is face-centered cubic (FCC), and the two have a large difference in their densities, melting points, and electronegativity. These pseudo alloys have exceptional properties such as high strength, good wear resistance, high arc erosion resistance, low thermal expansion coefficient as well as high thermal and electrical conductivity, thereby making them applicable for high-performance electrical contacts, heat sinks, plasma-facing materials in nuclear fusion reactors, electrodes for Electrical Discharge Machining (EDM), radiation shielding materials, and microwave communication systems [2–13]. Due to the high processing temperature of tungsten and mutual insolubility, it is difficult to process these pseudo-alloys using conventional methods, hence, techniques such as liquid infiltration and liquid phase sintering are employed to sinter them [14–16]. However, it is not easy to obtain full densification even through these processes.

Therefore, to ensure full densification, either high temperatures, high compaction pressure, longer holding time, or the addition of other transitions elements (namely Fe, Ni, or Co) are required. High temperatures or long holding time results in a non-homogeneous microstructure due to the copper leaching out. High compaction pressure limits the particle rearrangement due to increased particle–particle contact. The addition of Ni, Fe, and Co results in low electrical and thermal properties. From various experimental investigations, it has been found that the starting particle size and uniformity of the tungsten and copper powders significantly affect the magnitude of the sintering temperature and sinterability, and improve the final properties of the W–Cu composites [17–20].

Similarly, temperature and powder particle size are two main factors that determine sinterability. High-energy mechanical milling of tungsten and copper powders to produce fine or ultra-fine particles resulted in high relative density and lower sintering temperatures [3,20]. Processing techniques such as high-pressure sintering, microwave sintering, dynamic consolidation, and low-temperature infiltration in supergravity fields have resulted in higher density, improved mechanical properties, and a significant reduction in processing time [2,3,15,21–23]. As the copper content increased in the W–Cu composite, the electrical conductivity increased, as expected [21]. However, despite the higher copper content, the electrical conductivity reduced as a result of higher sintering temperature [20]. To overcome the drawbacks of liquid phase sintering, conventional sintering, and reduce the processing time and temperature, SPS. was employed in the current work to fabricate W–Cu composites at 900 °C. As the starting particle size affects the sintering temperature, as-received nano-copper powder was used in this study.

2. Experimental

2.1. Materials

Nano copper powder (Sigma-Aldrich, Bengaluru, India) with a particle size less than 100 nm and 99.99% purity and tungsten powder (Sigma-Aldrich, Bengaluru, India) with a particle size of ~12 μm were used as an as-received form. Figure 1a shows the powder morphology and size of the as-received tungsten powder as observed with the field emission scanning electron microscope, and Figure 1b shows nano-copper powder morphology and particle size as observed with a transmission electron microscope. The W–Cu (0, 5, 10, 15, 20, 25 wt.%) mixtures were prepared by weighing the required amounts of powders, followed by manual mixing in an agate mortar for 10 min and mixing in a planetary ball mill for 15 min in the dry state at 200 rpm without balls. The mixture was transferred into the SPS machine for sintering. The surface of the punch and die was separated from the powder mixture using a thin graphite sheet.

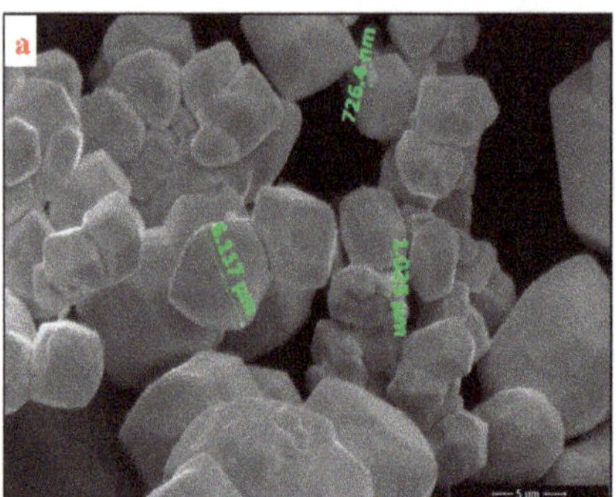
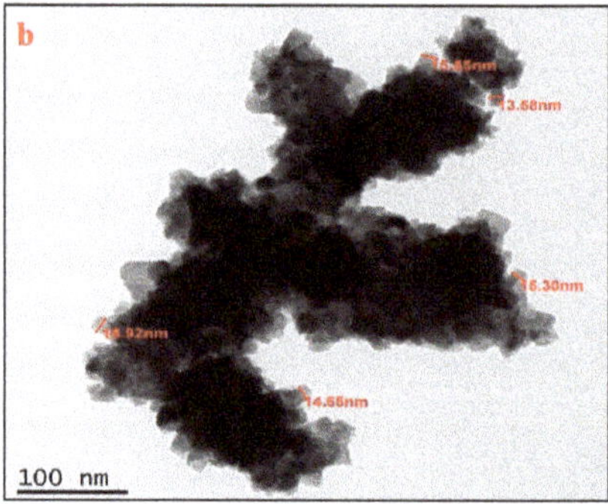

Figure 1. Powder morphology of as-received (**a**) tungsten powder and (**b**) nano-copper powder.

2.2. Spark Plasma Sintering (SPS)

SPS can be used to produce sintered samples rapidly in a single step using a combination of pressure, temperature, and electric field. In contrast to conventional sintering and other novel sintering techniques, SPS can significantly reduce the overall processing time through its high heating rates (about 100 °C per minute). Depending on the graphite die and punch geometry, very high heating rates such as 1000 °C per minute can also be attained. This is possible because of the electrical conductivity of the tool materials used in the SPS process. The low voltages applied across the set-up produce high currents, resulting in effective joule heating. Simultaneously, a uniaxial load is applied to the powder mixture to enhance the densification [24,25]. The sintering temperature of 900 °C in a vacuum with a heating rate of 100 °C per minute at a uniaxial pressure of 50 MPa in a graphite die of 30 mm diameter (model: Dr. Sinter 21050, supplier: Suji Electronic Industrial Co, LTD, Japan) was used. A voltage of 30 V and a current of 600 A was applied to process the compacts. A thin graphite sheet was used to separate the powder mixture from the surface of the die. Subsequently, after reaching the desired temperature of 900 °C, the samples were soaked for 120 s. After that, the furnace was turned off, and the compacts were left in the furnace to cool to room temperature.

2.3. Characterization of Sintered Samples

The density of the sintered samples was calculated using the Archimedes principle with three iterations for a sample. The samples were initially grounded using Emery sheets (grit size 220, 400, 600, 800, 1000, 1200, 1500, and 2000, respectively) and then polished to scratch-free surface using polishing machine (supplier: Chennai Metco Pvt. Ltd, Chennai, India) in the alumina media. According to the ASTM E407 standard [26], the polished samples were etched using Murakami's reagent containing 100 mL of water, 10 g of potassium hydroxide (supplier: Finox Pellets Industries, Gujarat, India), and 10 g of potassium ferricyanide (supplier: Hemadri Chemicals, Mumbai, India) to obtain the optical micrographs and thereby to determine the grain size of the samples. The line intercepts method was used on optical micrographs at 500x magnification. The samples' SEM images were obtained in both secondary electron mode and back-scattered electron mode (model: E.V.O. 18 Research, supplier: Zeiss, Oberkochen, Germany). X-ray powder diffractometry was used to analyze the sintered samples' phase composition (model: X'Pert^3 Powder, supplier: Malvern PANalytical, Malvern, UK) with a scan speed of 10 degrees per minute. Micro-tensile test specimens were machined from the sintered samples using wire-cut EDM and were tested with a crosshead speed of 1 mm/min (model: 8801, supplier: Instron, Norwood, MA, USA). The Vickers's micro-hardness of the samples was measured with a load of 0.1 kgf and with a dwell time of 10 s according to ASTM E92 [27] (model: MMT-X, supplier: Matsuzawa Co., Ltd, Japan). On each sample, twenty indentations were made randomly across the cross-section to obtain a mean value. The samples' electrical conductivity was measured according to ASTM E1004 [28] using the eddy current probe set up (Supplier: Technofour, India).

3. Results and Discussion

3.1. Density

The increase in *nano*-copper content in the compacts eventually led to increased relative density, with the highest being 82.26% for the C6 (refer to Table 1) compact. It is known that the increase in surface area increases the flow of powder particles. A combination of pressure and electric current enhanced the sinterability, even at a lower sintering temperature in the present work. Furthermore, the use of *nano*-copper powder facilitated the sinterability. The increase in relative sintered density of the compacts can be attributed to the flow of semi-solid *nano*-copper into the pores between the tungsten–tungsten grains. There was no substantial increase in the relative densities between the C5 and C6 compacts, which can be the effect of agglomeration of the particles. However, the relative density results of the compacts showed that the final stage of sintering, which results in pore shrinkage, was not entirely accomplished.

Table 1. Composition of the samples, relative density, and their respective notations.

Composition (wt.%).	Relative Density (%)	Notation
W100%	69.76	C1
W95%Cu5%	71.89	C2
W90%Cu10%	77.34	C3
W85%Cu15%	79.91	C4
W80%Cu20%	81.43	C5
W75%Cu25%.	82.26	C6

3.2. X-ray Diffraction (XRD)

XRD patterns of the sintered samples are shown in Figure 2. It was observed that the position and intensity of the diffraction peaks matched the reference plots [29], indicating the presence of only W and Cu phases. No oxide phase was found in the XRD patterns, implying the success of sintering in vacuum conditions. From the plot, it is evident that copper peaks were enhanced, which indicate good (semi-solid) flow of the nano-copper to fill the pores and thus obtain good densification.

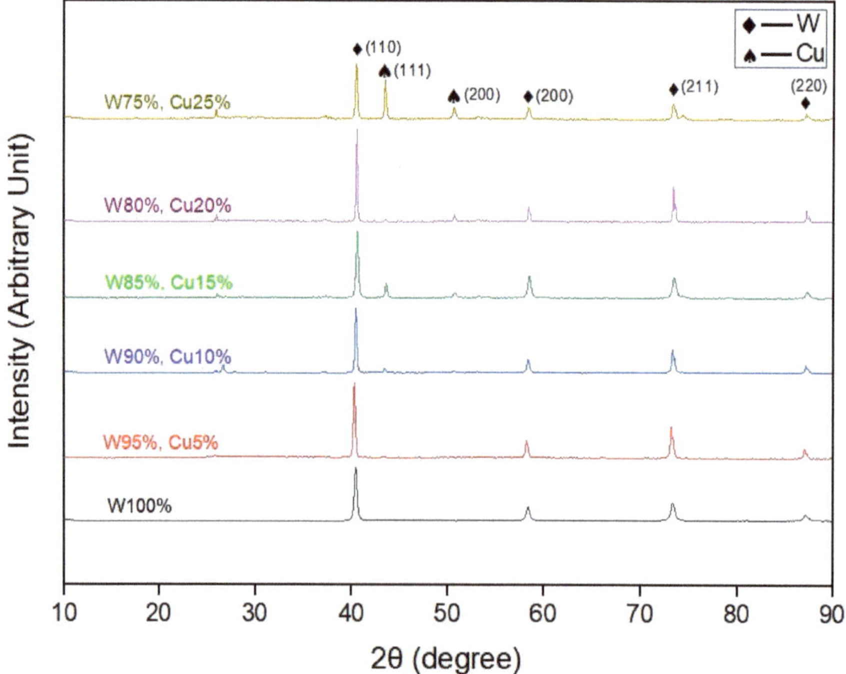

Figure 2. X-ray diffraction (XRD) profile of the sintered samples.

3.3. Microstructure

The optical micrographs of the sintered W–(*nano*)Cu compacts are shown in Figure 3. A two-phase microstructure with bright white tungsten grains embedded in a reddish-brown copper matrix was observed. In the compacts with lower *nano*-copper content (i.e., in C2, C3, and C4), an even distribution but a discontinuous network of copper phase was observed. In the C5 and C6 compacts, a more uniform distribution and continuous network of the matrix phase were observed, which also conformed to the elemental mapping shown in Figure 4. This might have contributed to the enhancement in the densification and reduction of porosity. Due to the low solubility of either element in one another, aggregation of *nano*-copper powder particles around the tungsten grains was observed. The compacts' grain size is listed in Table 2; the base tungsten compact (C1) had an average grain size of (20 ± 2) μm, which was reduced to (15 ± 1) μm of the C6 compact. This reduction in the compacts' grain size (C1 through C6) can be attributed to the increased *nano*-copper content, which inhibited the tungsten grain growth, resulting in relatively refined grains.

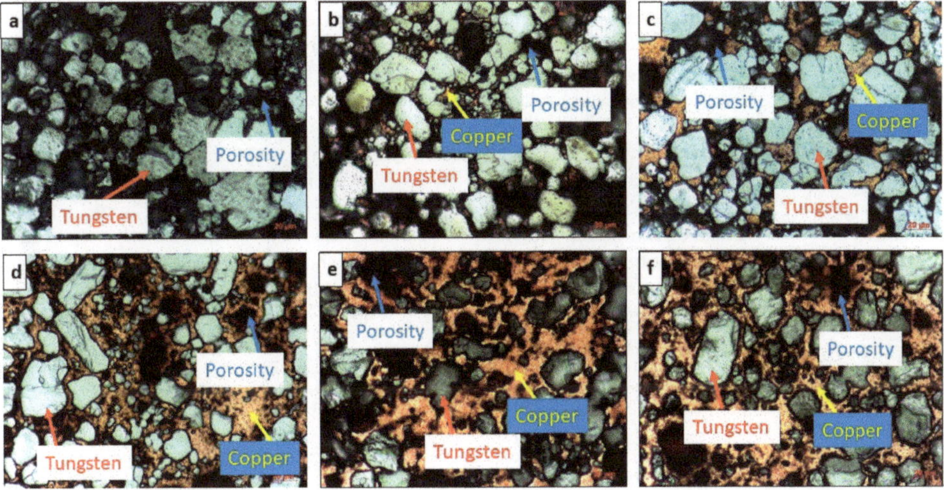

Figure 3. Optical micrographs of (**a**) W100%; (**b**) W95%, Cu5%; (**c**) W90%, Cu10%; (**d**) W85%, Cu15%; (**e**) W80%, Cu20%; (**f**) W75%, Cu25% compacts.

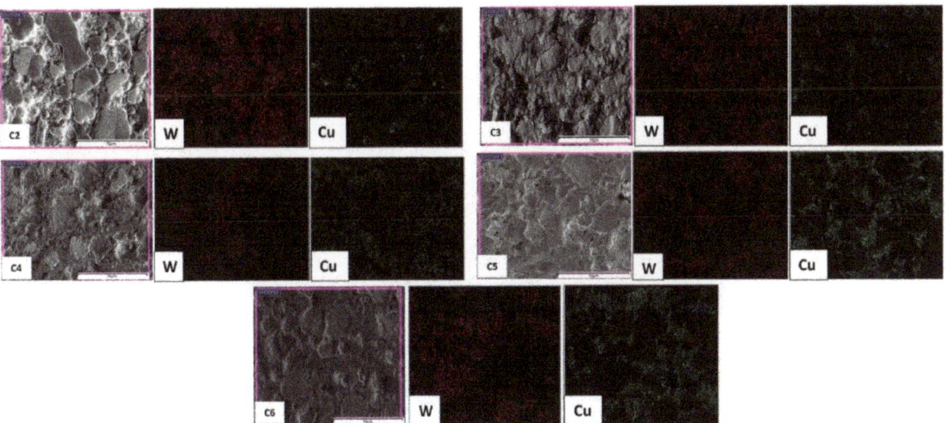

Figure 4. Elemental mapping of various compositions.

Table 2. Grain size (of tungsten).

Composition.	Grain Size (μm)
W100%	20 ± 2
W95%, Cu5%	19 ± 2
W90%, Cu10%	18 ± 1
W85%, Cu15%	16 ± 1
W80%, Cu20%	16 ± 1
W75%, Cu25%	15 ± 1

3.4. Mechanical Properties (Behavior)

To assess the mechanical properties, micro-tensile and micro-hardness tests were carried out, and the results are shown in Figures 5–8. The results show that the addition of *nano*-copper resulted in improved tensile strength and micro-hardness. Initially, there was no appreciable increase in strength or micro-hardness between samples C1 and C2. However, with the further increase in *nano*-copper content, significant improvement in the properties was observed, which was attributed to copper filling the pores and forming a network around the tungsten grains, thereby inhibiting grain growth and consequently increasing tensile strength (Figure 5) and reducing %elongation (Figure 7). From the Hall–Petch equation, it is well known that a decrease in grain size increases yield strength [30], which conforms to the obtained results, as shown in Figure 6. As a result, a nearly 50% increase in (tensile) strength and 37% increase in micro-hardness (Figure 8) were observed between samples C1 (W) and C6 (25% Cu). This increase can be attributed to both higher densification and grain refinement. The sample containing 25 wt.% *nano*-copper (C6) exhibited the highest tensile strength and micro-hardness values of 415 MPa and 341.44 $HV_{0.1}$, respectively.

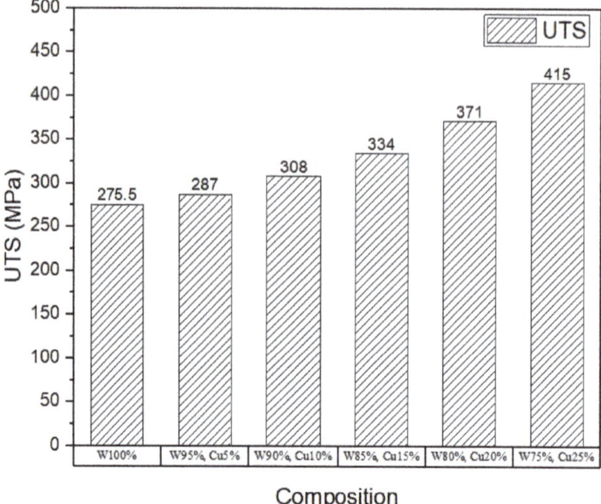

Figure 5. Plot of ultimate tensile strength.

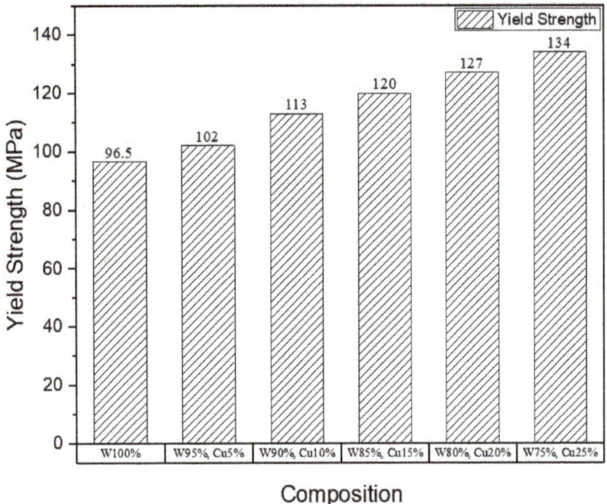

Figure 6. Plot of yield strength.

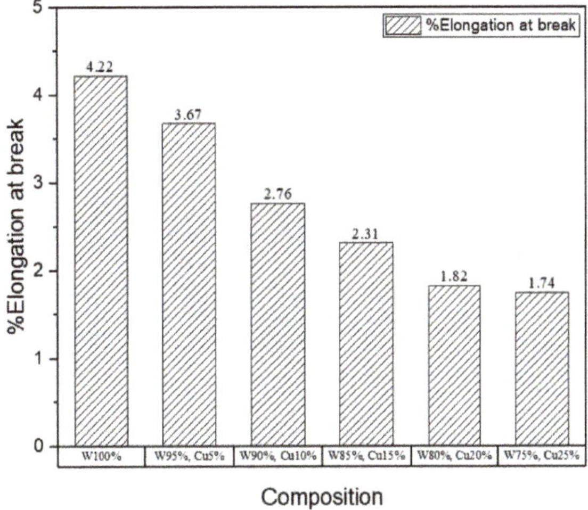

Figure 7. Plot of %elongation.

3.5. Fractography

Figure 9 shows the SEM images of the fractured surface of the specimens tested for tensile strength. As expected, the brittle fracture was observed in pure tungsten; the failure started by separation of W–W grains and propagating along the grain boundaries. The addition of *nano*-copper to tungsten resulted in a mixed-mode fracture. Due to low wettability and low diffusion coefficient between tungsten and copper at 900 °C, sufficient inter diffusion did not occur [31] and as a result, tungsten grains are pulled out of the copper matrix. The inter-granular fracture was observed in the regions with low copper content. Cleavage fracture occurred at coarse tungsten particles. The tungsten grain regions, which were well surrounded by copper matrix, failed in trans-granular mode [32]. Decohesion of tungsten particles was observed, as shown in Figure 10a. Due to the different deformation rates between hard tungsten particles and the ductile copper matrix, voids are formed. Eventually, these voids grew and were linked up, resulting in dimple type fracture. In Figure 10b, different modes of failure can be observed including the separation of tungsten particles from the copper matrix, dimple fracture, cleavage fracture, and tungsten grain being pulled out of a copper matrix.

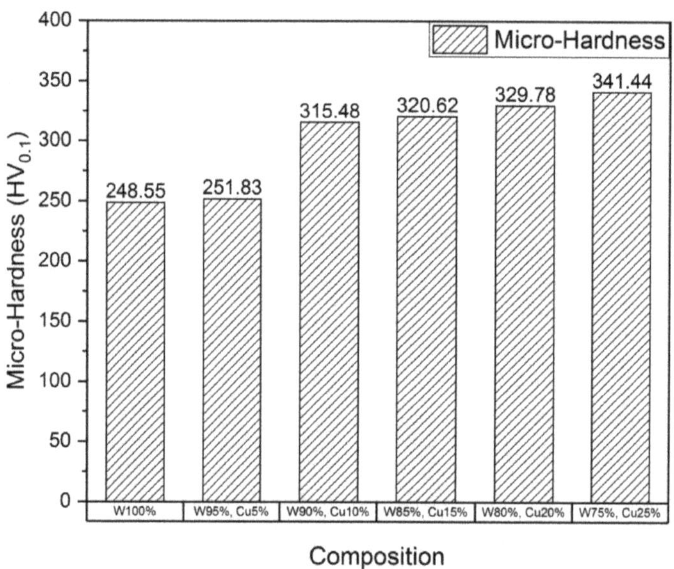

Figure 8. Plot of micro-hardness.

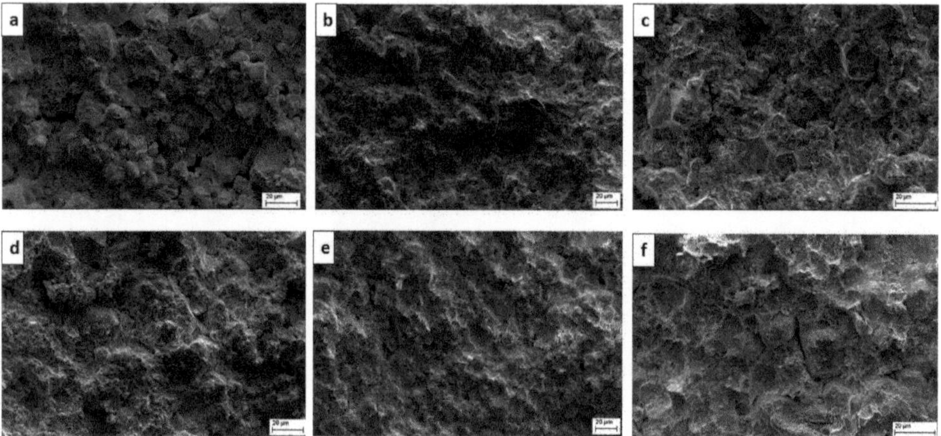

Figure 9. Scanning electron microscope (SEM) images of the fractured surface of samples tested for tensile strength. (**a**) W100%; (**b**) W95%, Cu5%; (**c**) W90%, Cu10%; (**d**) W85%, Cu15%; (**e**) W80%, Cu20%; (**f**) W75%, Cu25%.

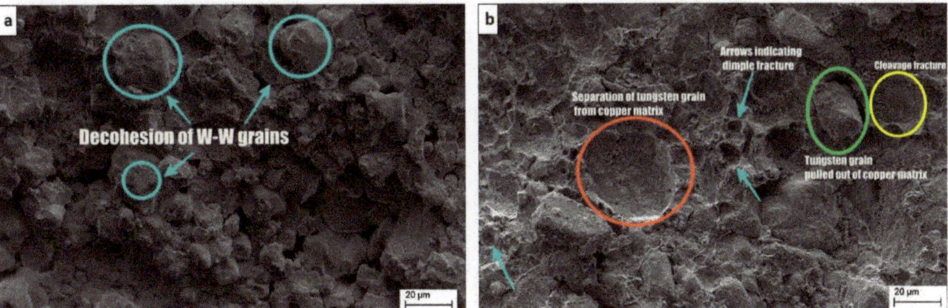

Figure 10. Image showing failure modes. (**a**) Base and (**b**) alloy with 25 wt.% nano-copper.

3.6. Electrical Conductivity

The eddy current probe method was employed to measure the samples' electrical conductivity, and the results are shown in Figure 11. The frequency of the alternating current is adjusted automatically by the device itself. As expected, increasing the *nano*-Cu content in samples resulted in increased electrical conductivity. However, the increase was not proportional to the increase in *nano*-Cu content. This can be attributed to the porosity of the samples, which restricts the flow of electrons. The highest electrical conductivity was found to be 28.2% IACS in the sample containing 25% *nano*-copper. The substantial increase in electrical conductivity can be attributed to the more uniform and continuous network of copper in the compact.

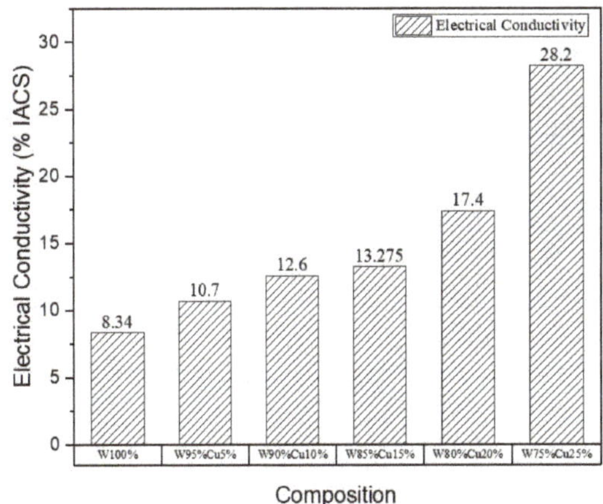

Figure 11. Electrical conductivity of the sintered samples.

4. Conclusions

From the present study, it can be concluded that W–(*nano*)Cu composites can be fabricated with high density using the spark plasma sintering technique even at 900 °C. Properties such as relative density, tensile strength, micro-hardness, and electrical conductivity were found to increase with nano copper addition. No intermetallics were formed, as confirmed by XRD analysis. As the copper content increased, a more uniform and continuous network of the matrix phase was observed, reducing porosity. The microstructure observed that the third stage of sintering, in which shrinkage or closure of pores occurs, was not wholly achieved under the processing conditions used in the present study.

However, an increase in uniaxial pressure or soaking time might overcome this and result in a further increase in relative density and other properties.

Author Contributions: Conceptualization, A.R.A. and A.M.; Methodology, A.R.A. and A.M.; Software, V.M. and M.S.; Validation, V.M. and M.S.; Formal analysis, A.R.A. and A.M. and C.-P.J.; Investigation, D.K.A., A.R.A. and A.M.; Resources, D.K.A., A.R.A., A.M. and C.-P.J.; Data curation, V.M. and M.S.; Writing, V.M. and A.R.A.; Writing—review and editing, V.M., A.M., A.R.A. and C.-P.J.; Visualization, A.R.A.; Supervision, A.R.A., A.M. and C.-P.J.; Project administration, A.R.A. and C.-P.J.; Funding acquisition, C.-P.J. All authors have read and agreed to the published version of the manuscript.

Funding: This research was funded by the Ministry of Science and Technology of the Republic of China (Taiwan), under grant numbers MOST 107-2221-E-194-024-MY3 and MOST 109-2923-E-194-002-MY3.

Institutional Review Board Statement: Not applicable.

Informed Consent Statement: Not applicable.

Data Availability Statement: The data that support the findings of this study are available from the corresponding author upon reasonable request.

Acknowledgments: The authors A. Raja Annamalai, acknowledge the Vellore Institute of Technology, Vellore campus in carrying out this research through RGEMS SEED grant.

Conflicts of Interest: The authors declare no conflict of interest.

References

1. Doré, F.; Lay, S.; Eustathopoulos, N.; Allibert, C.H. Segregation of Fe during the sintering of doped W–Cu alloys. *Scr. Mater.* **2003**, *49*, 237–242. [CrossRef]
2. Xu, L.; Yan, M.; Xia, Y.; Peng, J.; Li, W.; Zhang, L.; Liu, C.; Chen, G.; Li, Y. Influence of copper content on the property of Cu–W alloy prepared by microwave vacuum infiltration sintering. *J. Alloy. Compd.* **2014**, *592*, 202–206. [CrossRef]
3. Qiu, W.T.; Pang, Y.; Xiao, Z.; Li, Z. Preparation of W-Cu alloy with high Density and ultrafine grains by mechanical alloying and high-pressure sintering. *Int. J. Refract. Met. Hard Mater.* **2016**, *61*, 91–97. [CrossRef]
4. Chaubey, A.K.; Gupta, R.; Kumar, R.; Verma, B.; Kanpara, S.; Bathula, S.; Khirwadkar, S.S.; Dhar, A. Fabrication and characterization of W-Cu functionally graded material by spark plasma sintering process. *Fusion Eng. Des.* **2018**, *135*, 24–30. [CrossRef]
5. Zhu, X.; Cheng, J.; Chen, P.; Wei, B.; Gao, Y.; Gao, D. Preparation and characterization of nanosized W-Cu powders by a novel solution combustion and hydrogen reduction method. *J. Alloy. Compd.* **2019**, *793*, 352–359. [CrossRef]
6. Kwon, Y.S.; Chung, S.T.; Lee, S.; Noh, J.W.; Park, S.J.; German, R.M. Development of the High-Performance W-Cu Electrode. *Adv. Powder Metall. Part. Mater.* **2007**, *2*, 9.
7. Hiraoka, Y.; Inoue, T.; Hanado, H.; Akiyoshi, N. Ductile-to-brittle transition characteristics in W–Cu composites with an increase of Cu content. *Mater. Trans.* **2005**, *46*, 1663–1670. [CrossRef]
8. Ryu, S.S.; Kim, Y.D.; Moon, I.H. Dilatometric analysis on the sintering behavior of nanocrystalline W–Cu prepared by mechanical alloying. *J. Alloy. Compd.* **2002**, *335*, 233–240. [CrossRef]
9. Wang, W.S.; Hwang, K.S. The effect of tungsten particle size on the processing and properties of infiltrated W-Cu compacts. *Metall. Mater. Trans. A* **1998**, *29*, 1509–1516. [CrossRef]
10. Mordike, B.L.; Kaczmar, J.; Kielbinski, M.; Kainer, K.U. Effect of tungsten content on the properties and structure of cold extruded Cu-W composite materials. *PMI. Powder Metall. Int.* **1991**, *23*, 91–95.
11. Ho, P.W.; Li, Q.F.; Fuh, J.Y.H. Evaluation of W–Cu metal matrix composites produced by powder injection molding and liquid infiltration. *Mater. Sci. Eng. A* **2008**, *485*, 657–663. [CrossRef]
12. Amirjan, M.; Zangeneh-Madar, K.; Parvin, N. Evaluation of microstructure and contiguity of W/Cu composites prepared by coated tungsten powders. *Int. J. Refract. Met. Hard Mater.* **2009**, *27*, 729–733. [CrossRef]
13. Ibrahim, A.; Abdallah, M.; Mostafa, S.F.; Hegazy, A.A. An experimental investigation on the W–Cu composites. *Mater. Des.* **2009**, *30*, 1398–1403. [CrossRef]
14. Hamidi, A.G.; Arabi, H.; Rastegari, S. Tungsten–copper composite production by activated sintering and infiltration. *Int. J. Refract. Met. Hard Mater.* **2011**, *29*, 538–541. [CrossRef]
15. Mondal, A.; Upadhyaya, A.; Agrawal, D. Comparative study of densification and microstructural development in W–18Cu composites using microwave and conventional heating. *Mater. Res. Innov.* **2010**, *14*, 355–360. [CrossRef]
16. Pintsuk, G.; Brünings, S.E.; Döring, J.E.; Linke, J.; Smid, I.; Xue, L. Development of W/Cu—functionally graded materials. *Fusion Eng. Des.* **2003**, *66*, 237–240. [CrossRef]

17. Johnson, J.L.; Brezovsky, J.J.; German, R.M. Effects of tungsten particle size and copper content on densification of liquid-phase-sintered W-Cu. *Metall. Mater. Trans. A* **2005**, *36*, 2807–2814. [CrossRef]
18. Wei, X.; Tang, J.; Ye, N.; Zhuo, H. A novel preparation method for W–Cu composite powders. *J. Alloy. Compd.* **2016**, *661*, 471–475. [CrossRef]
19. Huang, L.M.; Luo, L.M.; Ding, X.Y.; Luo, G.N.; Zan, X.; Cheng, J.G.; Wu, Y.C. Effects of simplified pretreatment process on the morphology of W–Cu composite powder prepared by electroless plating and its sintering characterization. *Powder Technol.* **2014**, *258*, 216–221. [CrossRef]
20. Li, C.; Zhou, Y.; Xie, Y.; Zhou, D.; Zhang, D. Effects of milling time and sintering temperature on structural evolution, densification behavior, and properties of a W-20 wt. % Cu alloy. *J. Alloy. Compd.* **2018**, *731*, 537–545. [CrossRef]
21. Mondal, A.; Upadhyaya, A.; Agrawal, D. Effect of heating mode and copper content on the densification of W-Cu alloys. *Indian J. Mater. Sci.* **2013**, *2013*, 1–7. [CrossRef]
22. Zhanlei, W.; Huiping, W.; Zhonghua, H.; Hongyu, X.; Yifan, L. Dynamic consolidation of W-Cu nano-alloy and its performance as liner materials. *Rare Met. Mater. Eng.* **2014**, *43*, 1051–1055. [CrossRef]
23. Zhang, N.; Wang, Z.; Guo, L.; Meng, L.; Guo, Z. Rapid fabrication of W–Cu composites via low-temperature infiltration in supergravity fields. *J. Alloy. Compd.* **2019**, *809*, 151782. [CrossRef]
24. Guillon, O.; Gonzalez-Julian, J.; Dargatz, B.; Kessel, T.; Schierning, G.; Räthel, J.; Herrmann, M. Field-assisted sintering technology/spark plasma sintering: Mechanisms, materials, and technology developments. *Adv. Eng. Mater.* **2014**, *16*, 830–849. [CrossRef]
25. Suárez, M.; Fernández, A.; Menéndez, J.L.; Torrecillas, R.; Kessel, H.U.; Hennicke, J.; Kirchner, R.; Kessel, T. Challenges and opportunities for spark plasma sintering: A key technology for a new generation of materials. *Sinter. Appl.* **2013**, *13*, 319–342.
26. ASTM International. *Standard Practice for Microetching Metals and Alloys*; ASTM International E407-07; ASTM International: West Conshohocken, PA, USA, 2015.
27. ASTM International. *Standard Test Methods for Vickers Hardness and Knoop Hardness of Metallic Materials*; ASTM International E92; ASTM International: West Conshohocken, PA, USA, 2017.
28. ASTM International. *Standard Test Method for Determining Electrical Conductivity Using the Electromagnetic (Eddy Current) Method*; ASTM International E1004; ASTM International: West Conshohocken, PA, USA, 2017.
29. Liang, S.; Wang, X.; Wang, L.; Cao, W.; Fang, Z. Fabrication of CuW pseudo alloy by W–CuO nanopowders. *J. Alloy. Compd.* **2012**, *516*, 161–166. [CrossRef]
30. George, E.D. *Mechanical Metallurgy*, 2nd ed.; McGraw Hill Education: New York, NY, USA, 1976.
31. Chen, P.; Shen, Q.; Luo, G.; Wang, C.; Li, M.; Zhang, L.; Li, X.; Zhu, B. Effect of interface modification by Cu-coated W powders on the microstructure evolution and properties improvement for Cu–W composites. *Surf. Coat. Technol.* **2016**, *288*, 8–14. [CrossRef]
32. Liang, S.; Chen, L.; Yuan, Z.; Li, Y.; Zou, J.; Xiao, P.; Zhuo, L. Infiltrated W–Cu composites with the combined architecture of hierarchical particulate tungsten and tungsten fibers. *Mater. Charact.* **2015**, *110*, 33–38. [CrossRef]

Article

Study of Physico-Chemical Interactions during the Production of Silver Citrate Nanocomposites with Hemp Fiber

Alexandru Cocean [1], Iuliana Cocean [1], Georgiana Cocean [1,2], Cristina Postolachi [1], Daniela Angelica Pricop [3], Bogdanel Silvestru Munteanu [3], Nicanor Cimpoesu [1,4] and Silviu Gurlui [1,*]

[1] Atmosphere Optics, Spectroscopy and Laser Laboratory (LOASL), Faculty of Physics, Alexandru Ioan Cuza University of Iasi, 11 Carol I Bld, 700506 Iasi, Romania; alexcocean@yahoo.com (A.C.); iulianacocean@hotmail.com (I.C.); cocean.georgiana@yahoo.com (G.C.); tina.postolaki@gmail.com (C.P.); nicanornick@yahoo.com (N.C.)

[2] Rehabilitation Hospital Borsa, 1 Floare de Colt Street, 435200 Borsa, Romania

[3] Faculty of Physics, Alexandru Ioan Cuza University of Iasi, 11 Carol I Bld, 700506 Iasi, Romania; daniela.a.pricop@gmail.com (D.A.P.); msbogdanel18@yahoo.com (B.S.M.)

[4] Faculty of Material Science and Engineering, Gheorghe Asachi Technical University of Iasi, 59A Mangeron Bld, 700050 Iasi, Romania

* Correspondence: sgurlui@uaic.ro

Citation: Cocean, A.; Cocean, I.; Cocean, G.; Postolachi, C.; Pricop, D.A.; Munteanu, B.S.; Cimpoesu, N.; Gurlui, S. Study of Physico-Chemical Interactions during the Production of Silver Citrate Nanocomposites with Hemp Fiber. *Nanomaterials* **2021**, *11*, 2560. https://doi.org/10.3390/nano11102560

Academic Editor: Alexey Pestryakov

Received: 5 September 2021
Accepted: 26 September 2021
Published: 29 September 2021

Publisher's Note: MDPI stays neutral with regard to jurisdictional claims in published maps and institutional affiliations.

Abstract: In the study presented in this paper, the results obtained by producing nanocomposites consisting of a silver citrate thin layer deposited on hemp fiber surfaces are analyzed. Using the pulsed laser deposition (PLD) method applied to a silver target with impurities of nickel and iron, the formation of the silver citrate film is performed in various ways and the results are discussed based on Fourier Transform Infrared (FTIR) and Scanning Electron Microscopy coupled with Energy Dispersive X-ray (SEM-EDX) spectroscopy analyses. A mechanism of the physico-chemical processes that take place based on the FTIR vibrational modes and the elemental composition established by the SEM-EDS analysis is proposed. Inhibition of the fermentation process of Saccharomyces cerevisae is demonstrated for the nanocomposite material of the silver citrate thin layer, obtained by means of the PLD method, on hemp fabric. The usefulness of composite materials of this type can extend from sensors and optoelectronics to the medical fields of analysis and treatment.

Keywords: nanocomposites; pulsed laser deposition

1. Introduction

Composite materials, based on texturized textile fibers—as a reinforcing phase, on which silver films are deposited as a continuous phase (matrix)—are made for use in medical devices and also for other applications where an ionic state of silver (Ag^+) is required.

The aim of the experimental procedures presented in this work is to study if, during the pulsed laser deposition (PLD) process, the interaction of citric acid with the ablated silver, in its final plasma stage before hitting the support, and under vacuum chamber conditions, could lead to a chemical reaction that results in the formation of citrate. It is known that silver does not react with citric acid unless one of the reactants is in an ionic state (inorganic compounds of silver, such as $AgNO_3$, would enter into a reaction with citric acid, or sodium citrate would enter into a reaction with silver). Moreover, silver does not react with cellulose in its metallic state. In order to react with cellulose, "silver seeds", i.e., silver nitrate ($AgNO_3$), and reagents are required [1]. Other methods utilized for producing silver nanoparticle synthesis in an ionic state, for medical purposes, are based on so-called "green synthesis", where different biological organisms or extracts of those participate in a photochemical or chemical process [2–8]. Plasma reduction is another method used to synthesize Ag nanoparticles and Pt nanoparticles for further applications in dye-sensitized solar cells [9,10].

127

The impetus of this research, involving the production of compounds in which silver is in an ionic state, derives from its proven benefits regarding its antibacterial and antifungal properties. Researchers are also concerned with possible applications of gold nanoparticles in cancer treatment, where gold nanoparticles in an ionic state are supposedly the active form against cancer cells. In this respect, synthesis of gold nanoparticles has been achieved in the form of gold citrate but also as gold ionic bonded to polymers such as polyethylene imine [11]. Solar-driven water evaporation could be another application of such developed materials [12,13].

In the general context of interest for the study of noble metal nanoparticles, the method presented in this paper proposes both a new method of producing silver citrate embedded in a composite of the hemp-reinforcing phase, and a model of the physico-chemical mechanism of the process that takes place during, and under the conditions of, the pulsed laser deposition.

2. Materials and Methods

Pulsed laser deposition (PLD) was performed on the installation in the Atmosphere Optics, Spectroscopy and Lasers Laboratory [14] using the YG 981E/IR-10 laser system, with the parameters τ = 10 ns pulse width, λ = 532 nm wavelength, α = 45° incident angle and ν = 10 Hz pulse repetition time and $3\cdot10^{-2}$ Torr pressure, in the deposition chamber (Figure 1). The target subject for ablation (Figure 2) was made of silver with iron and nickel impurities. It was produced from jewelry scraps by means of mechano-thermic processes and chemical cleaning with 99.9% tetra borate ($Na_2B_4O_7\cdot10H_2O$) and 99% sodium borohydride ($NaBH_4$) to convert silver from the ionic to the atomic state [15,16] followed by treatment with baking soda as a catalyst ($NaHCO_3$ 99.5%) in the presence of aluminum foil in order to remove sulfur from the silver sulfide (Ag_2S).

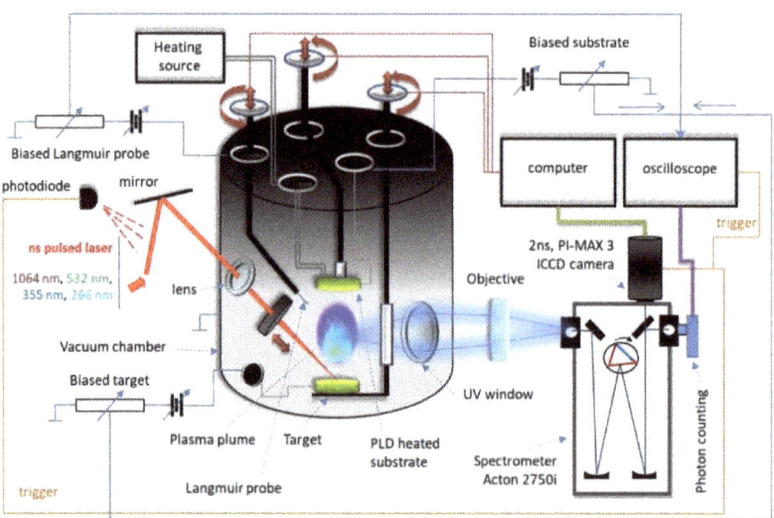

Figure 1. Experimental installation for PLD and LIBS in the Atmosphere Optics, Spectroscopy and Lasers Laboratory (http://spectroscopy.phys.uaic.ro. Accessed on 4 September 2021).

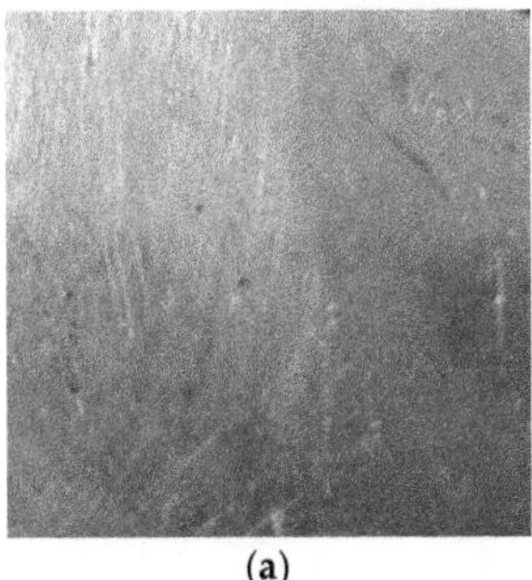 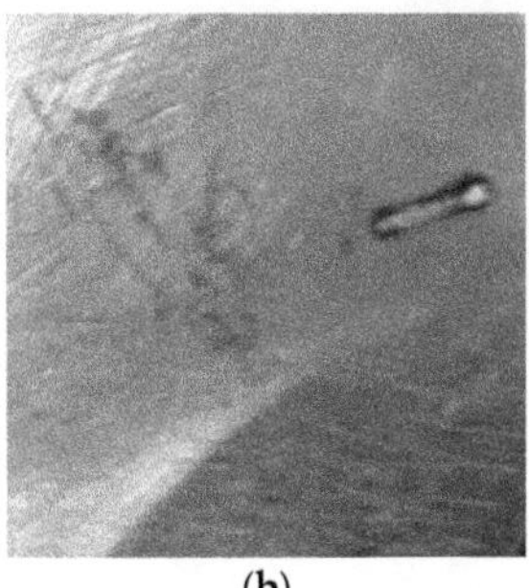

(a)   (b)

Figure 2. Silver target: (**a**) before irradiation; (**b**) after laser ablation during PLD processes.

The experimental studies refer to the physico-chemical interaction of plasma plume with the substrate containing citric acid. For that purpose, a total of three samples were produced, as follows:

- The film deposition on hemp fabric is noted as sample *D* (*Ag Film/Hemp*).
- The next film was deposited on a hemp fabric impregnated with supersaturated aqueous citric acid solution to form composite materials for further applications; this is referred to as sample *E* (*Ag Film/CA/Hemp*).
- Finally, silver film deposition was performed on a layer of citric acid applied as a supersaturated aqueous citric acid solution on a glass slab; this is denoted as sample *F* (*Ag Film/CA/Glass*).

The fabrics used as supports are noted as follows: HTND—the hemp twill fabric (natural); HTNS—the hemp twill fabric (natural) impregnated with citric acid solution.

Pulsed laser deposition was performed with laser energy of 150 mJ on a spot with a 168 μm average standard deviation, with a distance target support of 2 cm. The deposition time was 30 min for 18×10^4 pulses, the number of pulses necessary for obtaining a consistent silver layer [17].

A test to demonstrate the functional property of the new nanocomposite material, Ag-CA-HMP, related to inactivation of the microorganisms, was conducted using dry baker's yeast (Saccharomyces cerevisiae). Samples of hemp fabric, hemp fabric coated with a silver layer and hemp fabric coated with a silver citrate layer, obtained using the PLD method as described in the study presented herein, were placed on three glass slabs of 22 mm × 22 mm size, while one glass slab was used as a reference (R) or for the blind test. Each of the four glass slabs was placed in a different Petri dish. Quantities of 20 mg of yeast mixed with 10 mg of sugar were placed on the center of each sample and 0.5 mL of distilled water, at 45 °C–50 °C, was added on the mixture of yeast and sugar using a pipette. The samples obtained this way were named as follows:

- *1-Y-active* (yeast mixture on the glass slab or blind sample).
- *2-Y-HMP* (the yeast mixture on hemp fabric sample).
- *3-Y-Ag-HMP* (the yeast mixture on the silver layer deposited on the hemp fabric).
- *4-Y-Ag-CA-HMP* (the yeast mixture on the silver citrate layer deposited on the hemp fabric).

The reference or blind sample was noted as *R-Y-DRY*.

The surfaces and the foaming bubbles of CO_2 were measured in pixels using the Toup View software.

3. Results and Discussions

3.1. Target Initial Chemical Composition

After the preparatory steps of the target, EDX analysis showed a composition, in atomic percentages, of 81.84% Silver, 17.40% Nickel, and 0.76% iron in some areas; 88.26% Silver, 10.32% Nickel, 0.44% iron in other areas; and even 100% Silver in some areas. The composition demonstrated a non-homogenous distribution of the impurities through the silver matrix of the target, which is consistent with the usual conditions of manufacturing where materials of pure elements are not used due to their high cost and difficulties involved in purifying them, as well as the special storage conditions needed in order to avoid contamination/impurification.

3.2. Physico-Chemical Processes of Ablation and Redisposition on the Target

Analyzing the ablated area on the target after it was used in the PLD process, an important increase in iron and nickel was noticed compared with the results obtained before ablation. Thus, the elemental compositions of two areas, of 0.053 mm² each, that were analyzed at the center of the target after ablation (Figure 3b,c) were found as being of 64.56% Silver, 27.01% Nickel, 8.43% Iron and 77.20% Silver, and of 17.56% Nickel, 5.20% Iron, respectively. The detected elements were evidenced by the EDS spectrum presented in Figure 4. The increase in Ni and Fe atoms on the ablated area of the target was the result of different processes and phenomena during ablation such as re-deposition (Figure 3a–c) of the lighter elements (atomic mass, A, and atomic number, Z, of each component being $^{58.69}_{28}$Ni where $A_{Ni} = 58.69$, $^{55.845}_{26}$Fe where $A_{Fe} = 26$ and $^{108}_{47}$Ag where $A_{Ag} = 108$) that may have been due to their elastic collision with heavier atoms, ions and clusters, but also due to other perturbing phenomena including, but not limited to, the influence of electromagnetic fields on the charged particles (ions). Preferential Silver ablation was indicated with increase in Ni and Fe in the elemental composition of the ablated area, which is important for the quality of the deposited thin film.

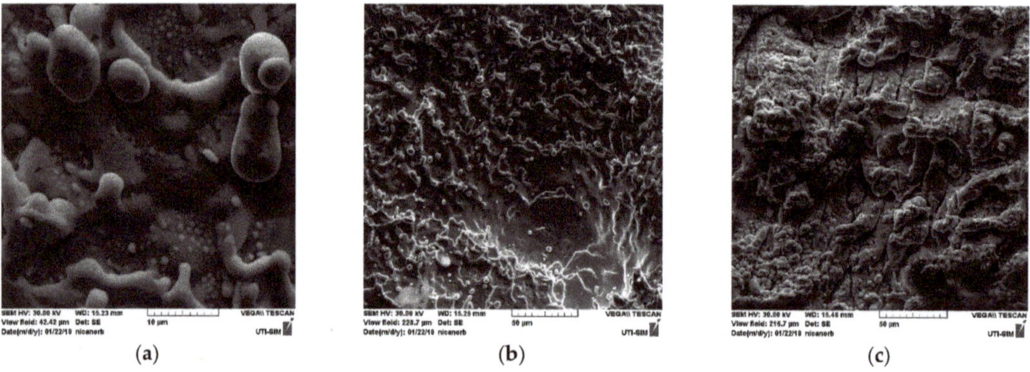

(a)　　　　　(b)　　　　　(c)

Figure 3. SEM images on ablated zone of the target with re-deposition structures: 5 kx magnitude (**a**); 1 kx magnitude (**b,c**).

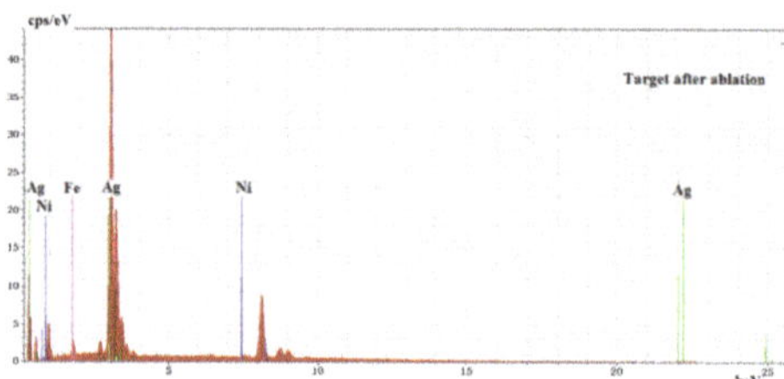

Figure 4. EDX spectrum of the target after ablation.

In the SEM images of Figure 3a–c, the droplets of the re-deposited material can be noticed on the ablated area.

3.3. Physico-Chemical Interaction of Silver Plume with the Citric Acid Substrate

The SEM images (500 × magnifying) of the three Silver film deposition samples are presented in Figure 5. The silver layer was not only deposited on the fabric surface. It penetrated through the interstices between the twisted fibers that formed the yarns from the fabric texture. In this way, the deposition produced a composite structure with new properties. The ionic state of the silver was the intended goal due to its antibacterial and antifungal properties and, for that reason, the study of the interaction of silver plasma plume with the citric acid from the support was important.

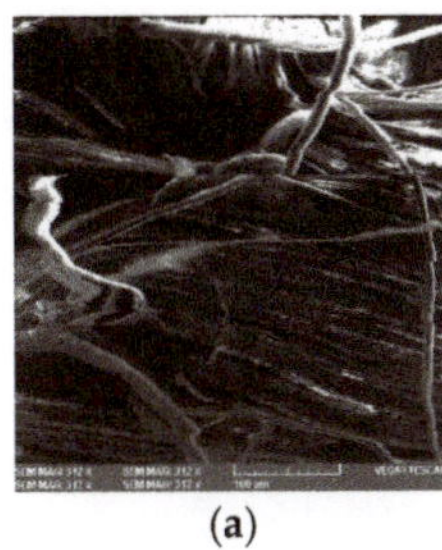
(a)

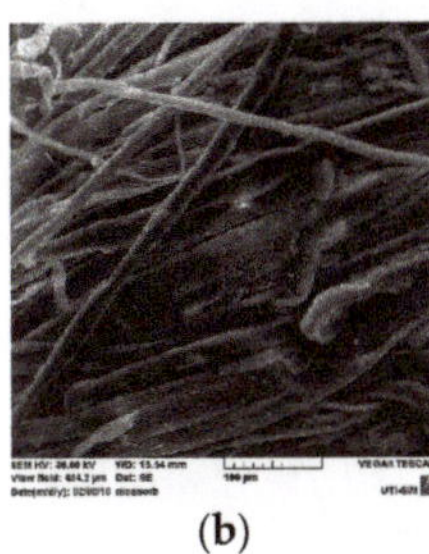
(b)

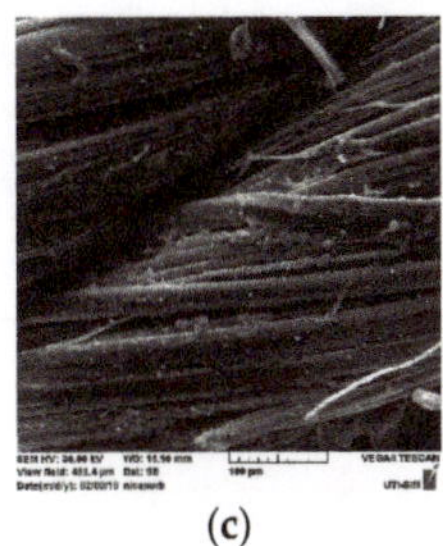
(c)

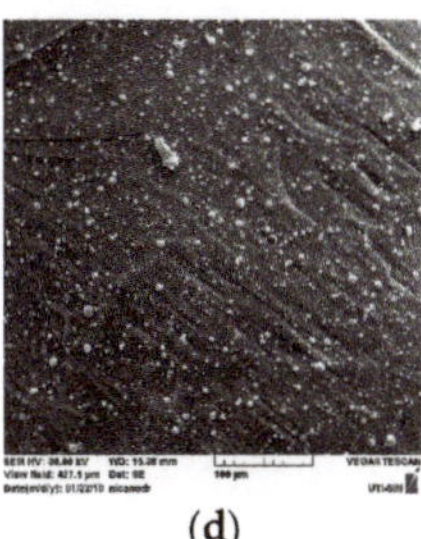
(d)

Figure 5. SEM images 500 x of the samples (**a**) Hemp fabric used as support for pulsed laser deposition; (**b**) Sample D (Ag Film/Hemp); (**c**) Sample E (Ag Film/CA/Hemp); (**d**) Sample F (Ag Film/CA/Glass).

The images in Figure 5 evidence the droplets of silver resulted during the laser deposition. EDS analysis showed an elemental composition on the thin layers as presented in Table 1 and in the spectra of Figure 6a–c. The analyzed areas were of 0.185 mm2 on the surfaces presented in Figure 5b–d) for data in Table 1.

Table 1. Elemental composition of the samples: Ag Film/Hemp (D); Ag Film/CA/Hemp (E); Ag Film/CA/Glass (F).

Element	Norm. wt.%			Norm. at.%		
	Sample (D) Ag Film/ Hemp	Sample (E) Ag Film/ CA/Hemp	Sample (F) Ag Film/ CA/Glass	Sample (D) Ag Film/ Hemp	Sample (E) Ag Film/ CA/Hemp	Sample (F) Ag Film/ CA/Glass
Oxygen	76.46	76.98	52.47	82.47	82.90	86.98
Silver	10.72	10.41	41.06	1.71	1.66	10.09
Carbon	10.58	10.33	-	15.20	14.82	-
Copper	2.23	2.28	-	0.60	0.62	-
Nickel	-	-	6.46	-	-	2.92
	100	100	100	100	100	100

In the elemental composition of the samples D and E, copper and carbon belong to the hemp fabric.

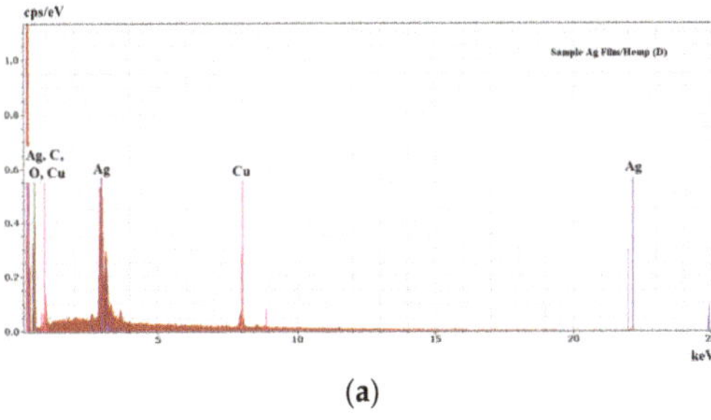
(a)

Figure 6. *Cont.*

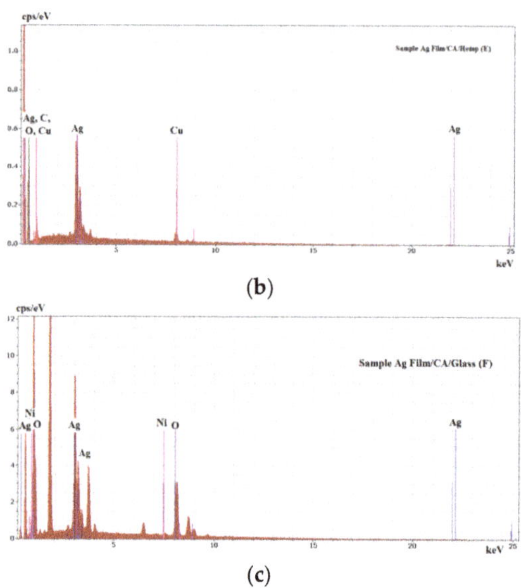

(b)

(c)

Figure 6. EDS spectra of the thin layers obtained by PLD: sample Ag Film/Hemp (D) (**a**); sample Ag Film/CA/Hemp (E) (**b**); sample Ag Film/CA/Glass (F) (**c**).

The PLD technique is based on laser ablation when the solid material of the target is transformed into liquid, gas and plasma. Plasma, also known as ionized gas [18–26], is a state of matter where ions, electrons, atoms and clusters exist at the same time. The predominant content in the Silver of the deposited thin film is noted in Table 1 This is in good accordance with the results of elemental composition of the ablated area on the target. It implies that the silver ions should arrive on the support surface and react with the citric acid. Because some of the silver was already derived from the ablation that occurred in liquid state, as per the temperature plots obtained from the COMSOL simulation of silver ablation and depositions shown in the work of A. Cocean et al., 2017 [27], the reaction rate was not expected to be high. Nevertheless, other phenomena can improve the active state of the film and increase its predisposition to subsequent reactions when in contact with biological agents such as bacteria and fungi. Regardless, this was not the experimental subject for this study, but it was where the idea started. Therefore, in order to determine whether the reaction took place, the powder of the deposited layers was collected by scraping it off the surface of the samples, and the FTIR results are presented in Figure 7. In order to avoid the possibility that silver ions could occur from oxidized areas or from Ag_2S that may form, in time, on the target surface when in contact with the atmosphere, borax was incorporated on the target (as presented in the Section 2). It followed that silver ions from oxides and/or sulfides had been converted into atoms before the PLD process, but also during PLD if any remaining oxygen traces were present in the deposition chamber. This provides further evidence that the only source of silver ions that will react with the citric acid is the plasma obtained via ablation during which ions, atoms and clusters co-exist. For a better evaluation of the FTIR spectra of the samples that were experimentally obtained, they were also compared with the spectra of the hemp fabrics used as supports for silver deposition, namely HTND (hemp fabric) and HTNS (hemp fabric impregnated with the supersaturated citric acid aqueous solution). The FTIR spectra (Figure 7) show the changes that citric acid undergoes as a result of silver deposition. The transformation of carboxyl groups into carboxylate ions can be observed in the FTIR spectra, and the formation of silver citrate is indicated.

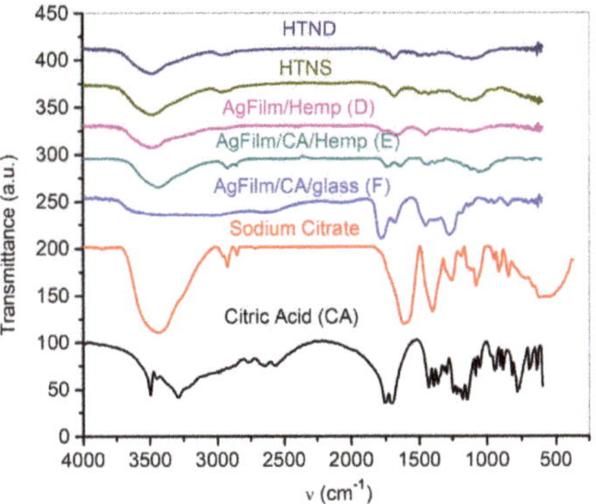

Figure 7. Comparison of the FTIR spectra of samples D (AgFilm/Hemp), E (AgFilm/CA/Hemp) and F (AgFilm/CA/glass), and the sodium citrate and citric acid (CA).

In this regard, for both types of hydroxyl groups assigned to carboxyl from citric acid (free: 3500 cm^{-1}; H-bounded: 3283 cm^{-1} [28,29]), there is evidence that they were transformed into ionized COO^{-} groups, which were involved in ionic bonds with Ag^{+}, as indicated by the carbonyl bands from 1638 cm^{-1} to 1760 cm^{-1} [28,29].

The bands from about 3500 cm^{-1} were the alcoholic groups [28,29] of the citric acid spectrum (CA), but they were also found in the sample of PLD film spectra for citrate when the peak became broader due to hydrogen bonding. This caused the silver ions from the plasma that was produced via ablation, that still existed when arriving on the substrate surface, to enter into a reaction with the citric acid, and thus, may have formed trisilver citrate as well as also mono- and disilver citrate (Figure 8). However, the physico-chemical process was more complex than only a citric reaction with silver plasma ions. The presence of intermolecular H-bonds between the hydroxyl groups of the citrate is indicated in the broad band at 3500 cm^{-1} in the sample E spectrum. Furthermore, it has been established that the partially ionized oxygen from the carbonyl groups will enter into hydrogen bonds with hydroxyl groups from other molecules, but also intramolecular bonds, as the peak from 1638 cm^{-1} could indicate [28,29]. H-bonding could also have occurred between the citrate and cellulose (Figure 9b). The spectrum of sample E (hemp impregnated with citric acid before deposition) evidences the formation of citrate. As for the silver that was deposited directly on the citric acid (sample F), the spectrum indicates that some of the citric acid had reacted with silver plasma ions, while some still existed as citric acid, or alternatively, only part of the carboxylic groups had been transformed into carboxylates (Figure 9c).

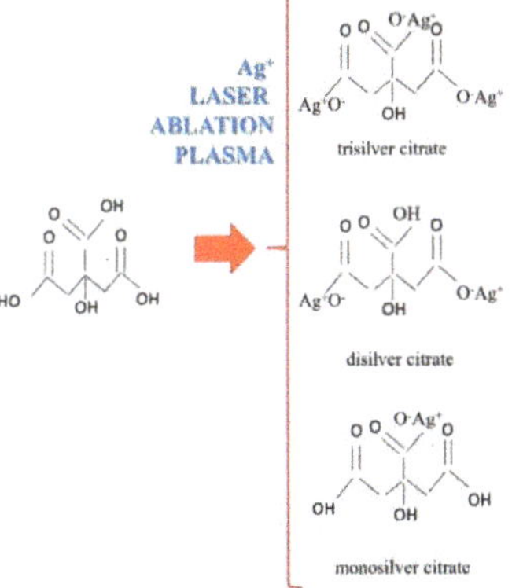

Figure 8. Chemical reaction of citric acid with silver ions from the laser ablation plasma plume.

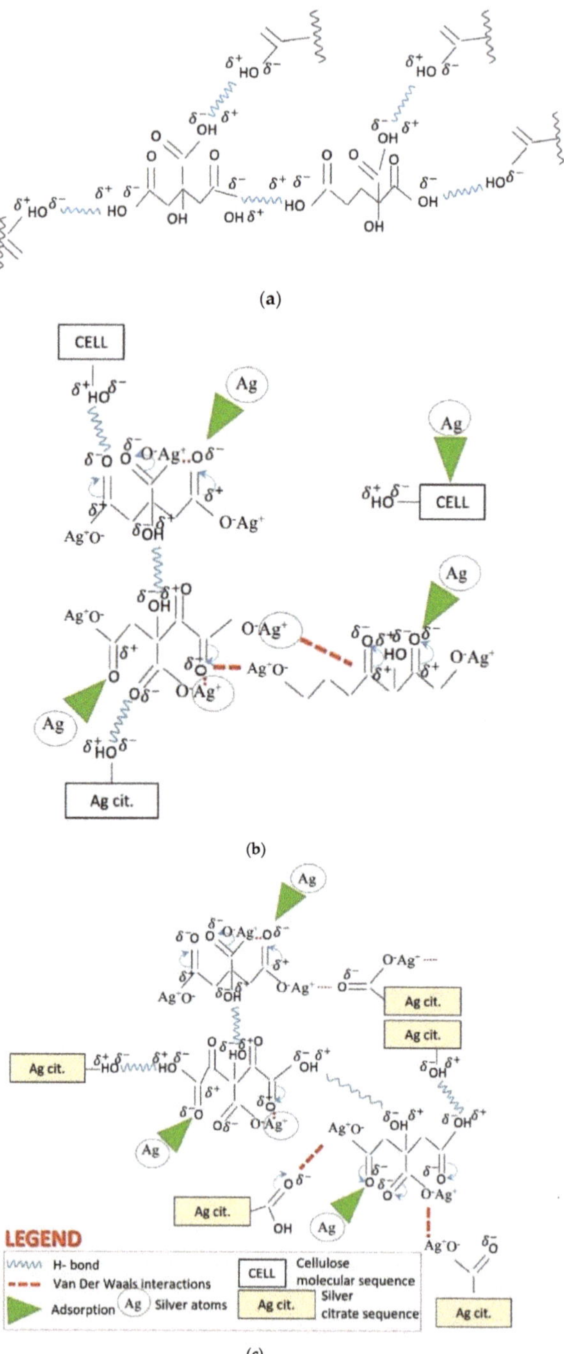

(a)

(b)

(c)

Figure 9. The reactions, namely the intermolecular and intramolecular interactions, that may occur upon the impact of silver plasma with the substrate surface containing citric acid: (**a**) hydrogen bonding in citric acid; (**b**) sequence of intermolecular H-bonding, inter- and intramolecular Van Der Waals interactions and Silver atoms' adsorption on citrate and cellulose; (**c**) sequence of intermolecular H-bonding, inter- and intramolecular Van Der Waals interactions and Silver atoms' adsorption on citrate.

Due to the environmental conditions in the vacuum chamber, it was possible for a further ionized structure of the silver citrate to form, which may have interacted through ionic and H-bonding and formed aggregates and complex structures, the most evident being for the citrate formed on the glass slab. The very broad band of the F (Ag Film/CA/Glass) spectra between 3560 and 2598 cm^{-1} indicates that both the carboxyl and carboxylate groups coexisted [28,29], meaning that part of the citric acid was transformed into citrate and part remained as citric acid. The broad band also indicates H-bonds, intermolecular ionic bonds and other intermolecular interactions, such as Van Der Waals interactions. Such a model may be depicted as in Figure 9a–c. Van Der Waals interactions between ionized and/or partially ionized atoms would have taken place in both films (the E-film on the hemp treated with citric acid and the F-film that was deposited on the citric acid). The adsorption of silver atoms on the carbonyl groups was also part of the complex interactions in the thin film system (Figure 9b,c).

The reactions, namely the intermolecular and intramolecular interactions, that may occur upon the impact of silver plasma with the substrate surface containing citric acid are schematically presented in Figures 8 and 9a–c. Further studies of this process of interaction between the silver plasma ions and the citric acid molecules during pulsed laser deposition, and its mechanism, may lead to a technique that can produce silver layers that are active against bacteria as well as being useful for other applications where an ionic state of silver and/or other metals is required. As noticed when analyzing the FTIR spectrum for silver deposited, by means of the PLD technique, on the citric acid, not all of the citric acid was transformed into citrate; this was because the quantity of the citric acid was excessive compared to the quantity of silver ions formed in the plasma. Based on this observation, a method to measure the quantity of the ions of metals that arrive on the substrate, and further, a PLD method of "titration", could be developed to analyze different intermediary compounds formed during the travelling of plasma on the path from the target to the substrate.

Regarding the sizes and shapes of the nanoparticles, UV-Vis spectral analysis was performed on sample F (Ag Film/CA/Glass). For the analysis, material was scraped from the upper part of the layer and was deposited on a glass slide (Ag Film/CA/Glass), with care taken not to scrape the glass directly. The material was dissolved in 2 mL of distilled and deionized water. The distilled and deionized water was used for comparison. Because the amount of material taken from the thin layer for UV-Vis analysis was very small and the dilution was high (imposed by the 2 mL cuvette volume), the signal obtained for absorption was of low intensity. However, peaks that, in the literature, were assigned to various sizes of silver nanoparticles, as well as silver citrate, can be observed. Thus, the UV-Vis spectrum (Figure 10) shows a succession of 284 nm, 353 nm, 412 nm, 443 nm peaks and a wide band between 487 nm and 543 nm. The peak at 284 nm could be related to the formation of $(Ag^+)_3$/citrate complexes [30], probably in the form of silver clusters and/or ultra-small silver nanoparticles [31]. Furthermore, in the same spectrum, a narrow band at 353 nm and a wide band between 487 and 543 nm appeared, possibly caused by the fusion of heated nanoparticles as a result of laser exposure. The spectral footprint of the fused nanoparticles is suggested by the wide band (between 487 and 543 nm) of the longitudinal oscillation mode of the surface plasmon and the narrow band at 353 nm of the transverse oscillation mode [32,33]. The fused nanoparticle formation could also be assigned to the broader bands, at 3500 cm^{-1}, of the FTIR spectra of samples E (AgFilm/CA/Hemp) and F(AgFilm/CA/Glass), but without neglecting other forms of aggregation, such as those described and presented in Figure 9.

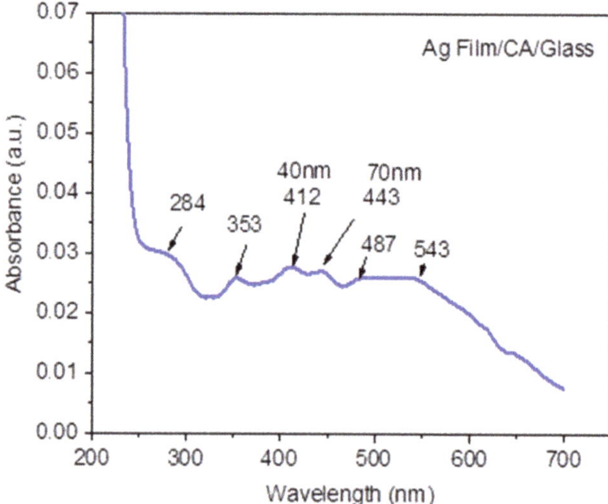

Figure 10. UV-Vis spectrum of sample F (Ag Film/CA/Glass).

Two other peaks at 412 nm and 443 nm can be attributed to the formation of triangular nanoparticles, with dimensions of around 40 nm [34], and nanoparticles with uneven shapes between 60 and 70 nm [35].

Citrate molecules act as both a capture ligand for silver particles and as a photoreducing agent for silver ions [36,37]. It is known that under the action of light radiation, citrate photoresists silver ions at the surface of previously formed silver seeds. The growth of nanoparticles with different morphologies is due to the rates of citrate reduction on certain faces of silver crystallites [34]. Silver–citrate complexes formed as a result of pulsed laser deposition indicate a possible application of the developed material as an antimicrobial layer [38].

3.4. Yeast Foaming Test for the New Ag-CA-HMP Material Synthesized by PLD Method

During the foaming test described in the Section 2, an intense foaming activity was observed on sample 1-Y from the first moment; the yeast on samples 2-Y-HMP and 3-Y-Ag-HMP was less active, while there was no foaming activity on sample 4-Y-Ag-CA-HMP (Figure 11).

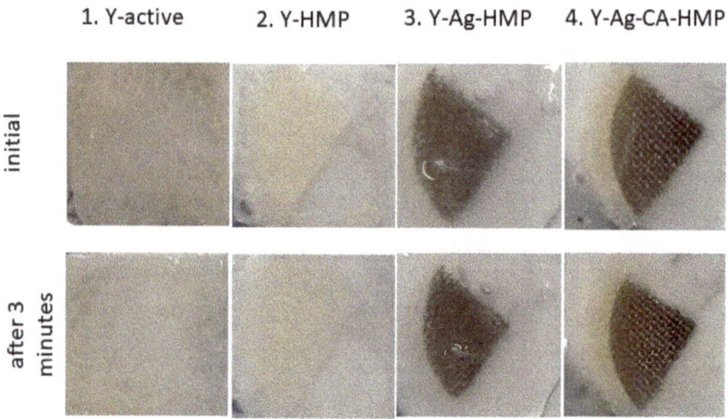

Figure 11. Images of foaming activity initial and after 3 min from starting the foaming test.

Based on the measurements of the samples and the dimensions of the glass slabs made using the Toup View software, in addition to the diameters of the CO_2 bubbles, the foaming activity (FA%) and foaming stability (FS%) were calculated for the initial moment of the foaming test, and at 3 min from the point at which the foaming test had started. The foaming activity and foaming stability were calculated as follows:

Glass slab area:

$$A = L \cdot l$$

The areas of samples 2, 3 and 4 (the samples on the fabric have one quarter of the area of the circles seen in Figure 11):

$$A = \frac{\pi \cdot R^2}{4} = \frac{\pi \cdot D^2}{16}$$

Areas of the bubbles:

$$a = \pi \cdot r^2 = \frac{\pi \cdot d^2}{4}$$

The total foam area on each sample as the sum of the areas of the bubbles:

$$F = \sum_{i=0}^{n} a_i$$

The foaming activity, as a percentage of the foam area from the sample area and from the slab area, respectively:

$$FA\% = \frac{F}{A} \cdot 100$$

The foaming stability, as a percentage of the foaming activity after 3 min from the point at which the foaming test was started, derived from the initial foaming activity of each sample:

$$FS\% = \frac{FA_{initial}(\%)}{FA_{after\ 3\ minutes}(\%)} \cdot 100$$

The results are presented in Table 2 and in Figure 12.

Table 2. Foaming activity (FA%) and foaming stability (FS%).

Sample	FA% (on Sample Surface)		FA% (on Slab Surface)		FS% (on Sample Surface)	FS% (on Slab Surface)
	Initial	after 3 min	Initial	after 3 min		
1	Y-active	89.42	9.71	89.42	9.71	12.23	12.23
2	Y-HMP	8.14	10.12	7.86	5.80	155.97	74.09
3	Y-Ag-HMP	7.37	6.99	3.21	2.80	86.86	74.18
4	Y-Ag-CA-HMP	0.00	0.00	0.34	0.00	-	0.00

The changes that occurred within 3 min of the initiation of foaming activity among the studied samples, using the surface of each sample and that of the glass slab on which the sample was placed as references, are presented in Figure 12a,b. The variations in foaming stability among the studied samples, using the area of each sample and that of the glass slab, are shown in Figure 13a,b.

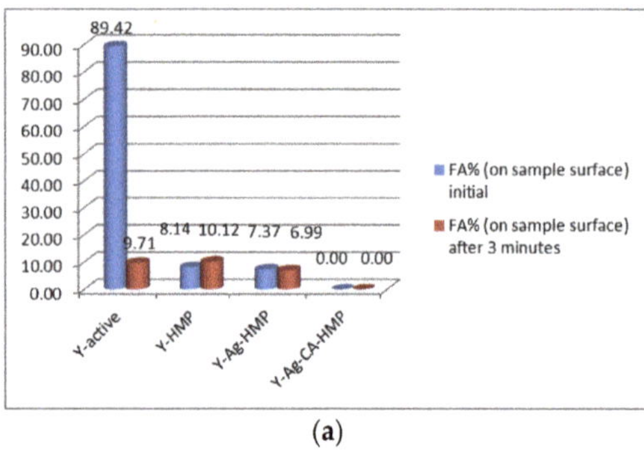
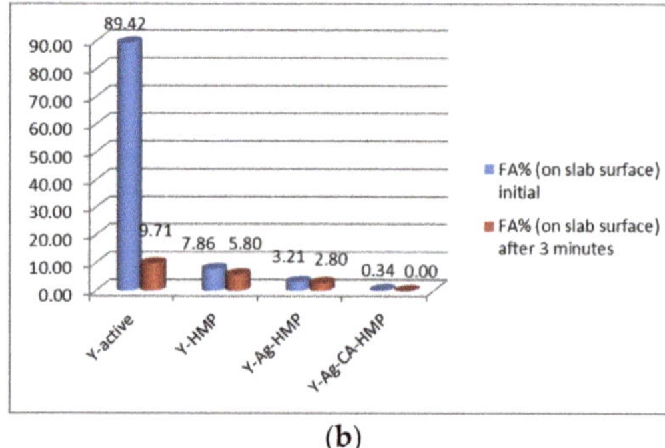

(a)   (b)

Figure 12. Comparison of yeast foaming activity (FA%) on the samples' surfaces (**a**) and on slab surface (**b**).

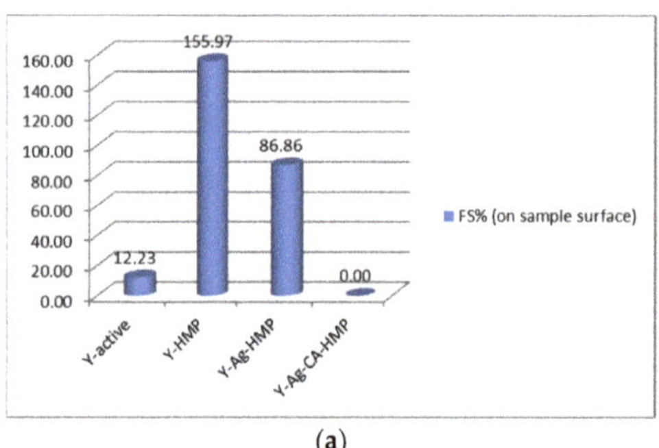
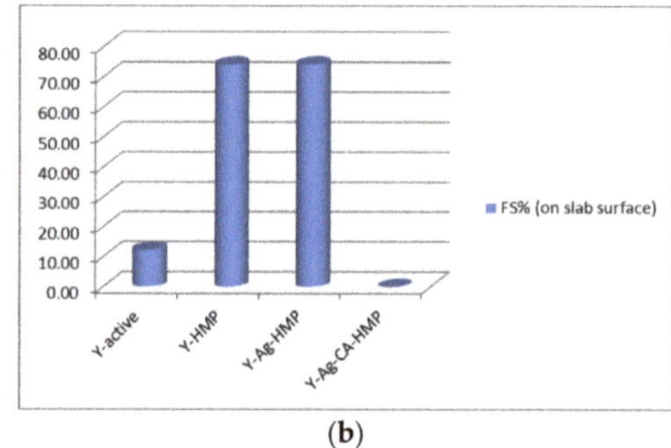

(a)   (b)

Figure 13. Comparison of yeast foaming stability (FS%) on samples' surfaces (**a**) and on slab surface (**b**).

The samples that were produced via the foaming test were dried at 25 °C for 24 h, and each sample was kept in a Petri dish. After drying, yeast material was collected from the glass slab of each sample. Each sample of yeast material was mixed with KBr in a mortar and, after that, was pressed into a ring and FTIR analysis was performed. The same process was undertaken with the sample of dry yeast that was used in the foaming test. The resulting spectra of the four samples and of the reference sample (dry yeast) are presented in Figure 14a,b.

The peaks that were specific to the FTIR spectrum of the dry yeast sample (Figure 14, Y-DRY) were also found in the spectra of the Y-active, Y-HMP and Y-Ag-HMP samples, with very small variation in terms of their intensity. The 3417 cm^{-1} band may be assigned to N-H stretching in the same range as the O-H-free and H-bonded samples. The bands assigned to proteins (the band at 1631 cm^{-1} assigned to CNO stretching and N-H bending in amides I, and the band at 1545 cm^{-1} assigned to C-N stretching and N-H bending in amides II) [39,40], as well as those assigned to sugars (1057 cm^{-1} and 908 cm^{-1}; these bands were specific to the cyclic ethers in carbohydrates) [39,40] or nucleic acids, denoted by the band at 1240 cm^{-1} as per A. Gallichet et al., 2001 [41], do not appear to have been essentially modified in the FTIR spectra of the mentioned samples.

Unlike the other three samples (Y-active, Y-HMP and Y-Ag-HMP), the test conducted the on Y-Ag-CA-HMP sample shows essential changes in the FTIR spectrum (Figure 14, Y-Ag-CA-HMP) compared to the spectrum of the dry yeast, and also in comparison to all other samples (Y-active, Y-HMP and Y-Ag-CA-HMP). In essence, the spectrum of the Y-Ag-CA-HMP sample shows the characteristics of sucrose (table sugar) even if some peaks appear to have overlapped those of yeast, which is also known to contain sugars.

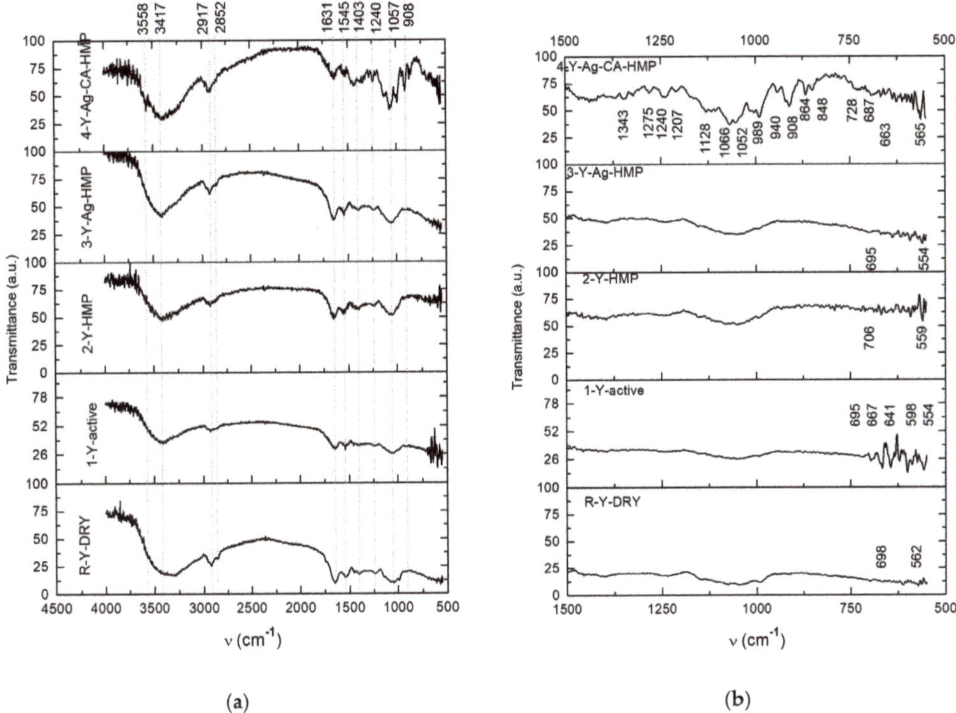

Figure 14. FTIR spectra of investigated behavior of yeast to different media provided by the analyzed materials (**a**) and detailed FTIR spectra in the finger-print region of 1500 cm^{-1}–500 cm^{-1} (**b**).

Of note is the strong and sharp vibrational band from 3558 cm^{-1}, which is specific to the spectrum of sucrose and is attributed to the free OH groups [41,42].

In the finger-print area of the Y-Ag-CA-HMP sample spectrum, the bands from 1403 cm^{-1}, 1343 cm^{-1}, 1275 cm^{-1}, 1240 cm^{-1}, 1207 cm^{-1}, 1128 cm^{-1}, 1066 cm^{-1}, 1052 cm^{-1}, 989 cm^{-1}, 940 cm^{-1}, 908 cm^{-1}, 864 cm^{-1}, 848 cm^{-1}, 728 cm^{-1}, 728 cm^{-1}, 680 cm^{-1} and 663–565 cm^{-1} are highlighted, and these are assigned to ethers (C–O–C stretch asymmetric; arC–O–alC) and to related components [14,28,29]; these bands can also be found in online databases on sucrose (1427 cm^{-1}, 1343 cm^{-1}, 1279 cm^{-1}, 1240 cm^{-1}, 1208 cm^{-1}, 1126 cm^{-1}, 1065 cm^{-1}, 1049 cm^{-1}, 988 cm^{-1}, 942 cm^{-1}, 908 cm^{-1}, 866 cm^{-1}, 849 cm^{-1}, 732 cm^{-1}, 687 cm^{-1} and 640–521 cm^{-1}) [39,40].

Based on the FTIR spectra shown in Figure 14, the interpretation of the foaming test results presented in Figures 12 and 13, as well as in Table 2, is that the fermentative effect of the yeast (Saccharomyces cerevisiae) on sugar was inactivated by the silver citrate layer that was obtained on the hemp fabric using the PLD technique and the method reported in this paper.

Thus, in the case of the three samples (Y-active, Y-HMP and Y-Ag-HMP), the fermentation of sugar is highlighted both by the CO_2 release process that can be observed in the foaming effect that occurred with the generation of the measured gas bubbles (Figures 12 and 13 and Table 2) and by the FTIR spectra, which was similar to that of the initial yeast spectrum. In the case of the Y-Ag-CA-HMP sample (the silver citrate layer deposited on the hemp fabric that we fabricated according to the new method described herein), no CO_2 bubbles were released during the foaming test, and in the immediate vicinity, there was only an insignificant amount of initial foaming activity (FA%) of 0.34%, with 0% foaming stability (FS%), at 3 min from initiation. This is in accordance with the

FTIR spectrum of sample Y-Ag-CA-HMP, which also shows that the fermentation of sugar did not take place.

4. Conclusions

The ionic state of silver is important for the production of composites with antibacterial and antifungal properties for various applications, including medical, but its importance is not limited to them. This study has shown that the PLD technique is suitable for the production of silver citrate as a result of the interaction of the ablation plume with a citric acid substrate or a substrate with citric acid contents. There are also indications resulting from the experiment presented herein that a method of "titration" may be developed in order to determine the quantity of metals ionic state in the plasma plume at different distances from the target, based on specific reactions. The method could be extended to other materials, with the aim of producing silver-based composites for later applications in industry and medicine, as well as for environmental investigations and air and water filtration. The tests and investigations show that the silver citrate that was obtained on the hemp fabric using our new method inactivated the yeast (Saccharomyces cerevisiae), confirming the effect of silver ions on the microorganisms, which, at the very least, inactivated their metabolism. This proves that a new functional material was obtained. Such nanocomposites of the silver citrate layer on the hemp fabric can be useful in processes where fermentation or excessive foaming induced by yeasts needs to be controlled by inhibition processes. Further investigations are needed and will be conducted to study the antimicrobial effect of the new nanocomposite material.

In order to determine the oxidation state of silver during deposition and on the deposited layer, in our further studies, we will consider the development of a method of analysis to be utilized in the deposition chamber. This method is necessary because the analysis would take place without the risk of contamination of the sample as a result of contact with air.

Author Contributions: Conceptualization, A.C., I.C., G.C., C.P. and S.G.; methodology, A.C., I.C., G.C., C.P. and S.G.; software, A.C.; validation, A.C., G.C., S.G., I.C.; formal analysis, G.C., A.C., I.C., S.G., C.P., D.A.P., B.S.M. and N.C.; investigation, A.C., G.C., I.C., S.G., B.S.M., N.C. and D.A.P.; resources, S.G.; data curation, A.C., I.C., G.C., C.P. and S.G.; writing—original draft preparation, A.C., I.C. and S.G.; writing—review and editing, A.C., G.C., I.C. and S.G.; visualization, S.G., A.C., and I.C.; supervision, S.G.; project administration, S.G.; funding acquisition, S.G. All authors have read and agreed to the published version of the manuscript.

Funding: This research was funded by the Ministry of Research, Innovation and Digitization, project FAIR_09/24.11.2020, and by the Executive Agency for Higher Education, Research, Development and Innovation, UEFISCDI, ROBIM, project number PN-III-P4-ID-PCE2020-0332.

Data Availability Statement: Not applicable.

Conflicts of Interest: The authors declare no conflict of interest.

References

1. Zeng, J.; Tao, J.; Li, W.; Grant, J.; Wang, P.; Zhu, Y.; Xia, Y. A Mechanistic Study on the Formation of Silver Nanoplates in the Presence of Silver Seeds and Citric Acid or Citrate Ions. *Chem.-Asian J.* **2010**, *6*, 376–379. [CrossRef]
2. Bera, R.K.; Raj, C.R. A facile photochemical route for the synthesis of triangular Ag nanoplates and colorimetric sensing of H_2O_2. *J. Photochem. Photobiol. A: Chem.* **2013**, *270*, 1–6. [CrossRef]
3. Butch, C.; Cope, E.D.; Pollet, P.; Gelbaum, L.; Krishnamurthy, R.; Liotta, C.L. Production of Tartrates by Cyanide-Mediated Dimerization of Glyoxylate: A Potential Abiotic Pathway to the Citric Acid Cycle. *J. Am. Chem. Soc.* **2013**, *135*, 13440–13445. [CrossRef]
4. Homan, K.A.; Souza, M.; Truby, R.; Luke, G.P.; Green, C.; Vreeland, E.; Emelianov, S. Silver Nanoplate Contrast Agents for In Vivo Molecular Photoacoustic Imaging. *ACS Nano* **2012**, *6*, 641–650. [CrossRef]
5. Zhang, L.V.; Brunet, P.; Eggers, J.; Deegan, R.D. Wavelength selection in the crown splash. *Phys. Fluids* **2010**, *22*, 122105. [CrossRef]
6. Jyoti, K.; Baunthiyal, M.; Singh, A. Characterization of silver nanoparticles synthesized using Urtica dioica Linn. leaves and their synergistic effects with antibiotics. *J. Radiat. Res. Appl. Sci.* **2016**, *9*, 217–227. [CrossRef]

7. Bagherzade, G.; Tavakoli, M.M.; Namaei, M.H. Green synthesis of silver nanoparticles using aqueous extract of saffron (*Crocus sativus* L.) wastages and its antibacterial activity against six bacteria. *Asian Pac. J. Trop. Biomed.* **2017**, *7*, 227–233. [CrossRef]

8. Karthik, L.; Kumar, G.; Kirthi, A.V.; Rahuman, A.A.; Rao, K.V.B. *Streptomyces* sp. LK3 mediated synthesis of silver nanoparticles and its biomedical application. *Bioprocess Biosyst. Eng.* **2013**, *37*, 261–267. [CrossRef] [PubMed]

9. Dao, V.-D.; Choi, H.-S. Highly-Efficient Plasmon-Enhanced Dye-Sensitized Solar Cells Created by Means of Dry Plasma Reduction. *Nanomaterials* **2016**, *6*, 70. [CrossRef]

10. Dao, V.-D.; Tran, C.Q.; Ko, S.-H.; Choi, H.-S. Dry plasma reduction to synthesize supported platinum nanoparticles for flexible dye-sensitized solar cells. *J. Mater. Chem. A* **2013**, *1*, 4436. [CrossRef]

11. Mohan, J.C.; Praveen, G.; Chennazhi, K.; Jayakumar, R.; Nair, S. Functionalised gold nanoparticles for selective induction ofin vitroapoptosis among human cancer cell lines. *J. Exp. Nanosci.* **2013**, *8*, 32–45. [CrossRef]

12. Dao, V.-D.; Vu, N.H.; Yun, S. Recent advances and challenges for solar-driven water evaporation system toward applications. *Nano Energy* **2019**, *68*, 104324. [CrossRef]

13. Dao, V.-D.; Vu, N.H.; Dang, H.-L.T.; Yun, S. Recent advances and challenges for water evaporation-induced electricity toward applications. *Nano Energy* **2021**, *85*, 105979. [CrossRef]

14. Cocean, I.; Cocean, A.; Postolachi, C.; Pohoata, V.; Cimpoesu, N.; Bulai, G.; Iacomi, F.; Gurlui, S. Alpha keratin amino acids BEHVIOR under high FLUENCE laser interaction. Medical applications. *Appl. Surf. Sci.* **2019**, *488*, 418–426. [CrossRef]

15. Cook, M.M.; Lander, J.A. Use of Sodium Borohydride to Control Heavy Metal Discharge in the Photographic Industry. *J. Appl. Photogr. Eng.* **1979**, *5*, 144–147.

16. Cotton, F.A.; Wilkinson, G. *Advanced Inorganic Chemistry*, 4th ed.; Interscience: New York, NY, USA, 1980.

17. Alonso, J.; Diamant, R.; Castillo, P.; Acosta–García, M.; Batina, N.; Haro-Poniatowski, E. Thin films of silver nanoparticles deposited in vacuum by pulsed laser ablation using a YAG:Nd laser. *Appl. Surf. Sci.* **2009**, *255*, 4933–4937. [CrossRef]

18. Gurlui, S.; Pompilian, G.O.; Nemec, P.; Nazabal, V.; Ziskind, M.; Focsa, C. Plasma Diagnostics in Pulsed Laser Deposition of GaLaS Chalcogenides. *Appl. Surf. Sci.* **2013**, *278*, 352–356.

19. Bechtel, J.H. Heating of solid targets with laser pulses. *J. Appl. Phys.* **1975**, *46*, 1585–1593. [CrossRef]

20. Kuznetsov, I.A.; Garaeva, M.Y.; Mamichev, D.A.; Grishchenko, Y.V.; Zanaveskin, M.L. Formation of ultrasmooth thin silver films by pulsed laser deposition. *Crystallogr. Rep.* **2013**, *58*, 739–742. [CrossRef]

21. König, J.; Nolte, S.; Tünnermann, A. Plasma evolution during metal ablation with ultrashort laser pulses. *Opt. Express* **2005**, *13*, 10598. [CrossRef]

22. Stafe, M.; Negutu, C.; Puscas, N.N.; Popescu, I.M. Pulsed laser ablation of solids. *Rom. Rep. Phys.* **2010**, *62*, 758–770.

23. Mercadier, L.; Hermann, J.; Grisolia, C.; Semerok, A. Analysis of deposited layers on plasma facing components by laser-induced breakdown spectroscopy: Towards ITER tritium inventory diagnostics. *J. Nucl. Mater.* **2011**, *415*, S1187–S1190. [CrossRef]

24. Stafe, M.; Vladoiu, I.; Negutu, C.; Popescu, I.M. Experimental investigation of the nanosecond laser ablation rate of aluminum. *Rom. Rep. Phys.* **2008**, *60*, 789–796.

25. Gurlui, S.; Sanduloviciu, M.; Strat, M.; Strat, G.; Mihesan, C.; Ziskind, M.; Focsa, C. Dynamic space charge structures in high fluence laser ablation plumes. *J. Optoelectron. Adv. Mater.* **2006**, *8*, 148–151.

26. Gurlui, S.; Agop, M.; Nica, P.; Ziskind, M.; Focsa, C. Experimental and Theoretical Investigations of a Laser Produced Aluminum. *Plasma Phys. Rev. E* **2008**, *78*, 026405. [CrossRef]

27. Cocean, A.; Cocean, I.; Gurlui, S.; Iacomi, F. Study of the pulsed laser deposition phenomena by means of Comsol Multiphysics. *Univ. Politeh. Buchar. Sci. Bull. Ser. A Appl. Math. Phys.* **2017**, *79*, 263–274.

28. Pretsch, E.; Buhlmann, P.; Badertscher, M. *Structure Determination of Organic Compounds*, 4th ed.; Revised and Enlarged Edition; Springer: Berlin/Heidelberg, Germany, 2009. [CrossRef]

29. Miller, F.A.; Wilkins, C.H. Infrared Spectra and Characteristic Frequencies of Inorganic Ions. *Anal. Chem.* **1952**, *24*, 1253–1294. [CrossRef]

30. Durdureanu-Angheluta, A.; Mihesan, C.; Doroftei, F.; Dascalu, A.; Ursu, L.; Velegrakis, M.; Pin-teala, M. Formation by laser ablation in liquid (lal) and characterization of citric acid-coated iron oxide nanoparticles. *Rev. Roum. Chim.* **2014**, *59*, 151–159.

31. Galazzi, R.; Santos, E.; Caurin, T.; Pessoa, G.; Mazali, I.; Arruda, M.A.Z. The importance of evaluating the real metal concentration in nanoparticles post-synthesis for their applications: A case-study using silver nanoparticles. *Talanta* **2016**, *146*, 795–800. [CrossRef]

32. Li, Y.; Ye, Y.; Fan, Y.; Zhou, J.; Jia, L.; Tang, B.; Wang, X. Silver Nanoprism-Loaded Eggshell Membrane: A Facile Platform for In Situ SERS Monitoring of Catalytic Reactions. *Crystals* **2017**, *7*, 45. [CrossRef]

33. Choi, B.-H.; Lee, H.-H.; Jin, S.; Chun, S.; Kim, S.-H. Characterization of the optical properties of silver nanoparticle films. *Nanotechnology* **2007**, *18*, 075706. [CrossRef]

34. Jia, H.; Xu, W.; An, J.; Li, D.; Zhao, B. A simple method to synthesize triangular silver nanoparticles by light irradiation. *Spectrochim. Acta* **2006**, *64*, 956–960. [CrossRef]

35. Paramelle, D.; Sadovoy, A.; Gorelik, S.; Free, P.; Hobley, J.; Fernig, D. A rapid method to estimate the concentration of citrate capped silver nanoparticles from UV-visible light spectra. *Analyst* **2014**, *139*, 4855–4861. [CrossRef]

36. Maillard, M.; Huang, P.; Brus, L. Nano-Silver Nanodisk Growth by Surface Plasmon Enhanced Photoreduction of Adsorbed [Ag^+]. *Nano Lett.* **2003**, *3*, 1611–1615. [CrossRef]

37. Ahern, A.M.; Garrell, R.L. In situ photoreduced silver nitrate as a substrate for surface-enhanced Raman spectroscopy. *Anal. Chem.* **1987**, *59*, 2813–2816. [CrossRef]
38. Djokić, S. Synthesis and Antimicrobial Activity of Silver Citrate Complexes. *Bioinorg. Chem. Appl.* **2008**, *2008*, 436458. [CrossRef]
39. Molon, M.; Zebrowski, J. Phylogenetic relationship and FTIR spectroscopy-derived lipid determinants of lifespan parameters in the Saccharomyces cerevisiae yeast. *FEMS Yeast Res.* **2017**, *17*. [CrossRef] [PubMed]
40. Galichet, A.; Sockalingum, G.D.; Belarbi, A.; Manfait, M. FTIR spectroscopic analysis of Saccharomyces cerevisiae cell walls: Study of an anomalous strain exhibiting a pink-colored cell phenotype. *FEMS Microbiol. Lett.* **2001**, *197*, 179–186. [CrossRef] [PubMed]
41. Sucrose. Available online: https://webbook.nist.gov/cgi/cbook.cgi?ID=C57501&Type=IR-SPEC&Index=0 (accessed on 4 September 2021).
42. ATR-FT-IR Spectrum of Sucrose (4000–225 cm^{-1}). Available online: https://spectra.chem.ut.ee/paint/binders/sucrose/ (accessed on 4 September 2021).

MDPI
St. Alban-Anlage 66
4052 Basel
Switzerland
Tel. +41 61 683 77 34
Fax +41 61 302 89 18
www.mdpi.com

Nanomaterials Editorial Office
E-mail: nanomaterials@mdpi.com
www.mdpi.com/journal/nanomaterials

9 783036 522470